The Intellectual Commons

Toward an Ecology of Intellectual Property

Henry C. Mitchell

LEXINGTON BOOKS

A Division of
ROWMAN & LITTLEFIELD PUBLISHERS, INC.
Lanham • Boulder • New York • Toronto • Oxford

LEXINGTON BOOKS

A division of Rowman & Littlefield Publishers, Inc.
A wholly owned subsidary of The Rowman & Littlefield Publishing Group, Inc.
4501 Forbes Boulevard, Suite 200
Lanham, MD 20706

PO Box 317, Oxford, OX2 9RU, UK

British Library Cataloguing in Publication Information Available

**The Library of Congress catalogued the hardcover edition of
this book as follows:**

Mitchell, Henry C., 1954-
    The intellectual commons : toward an ecology of
intellectual property / by Henry C. Mitchell, Jr.
        p. cm. — (Lexington studies in social, political,
and legal philosophy)
    Includes bibliographic references and index.
    ISBN 0-7391-0948-0 (cloth : alk. paper)
    1. Intellectual property—Philosophy. I. Title. II.
Series.
K1500.M58 2005
346.04'8'09—dc22   2004027142

    ISBN 0-7391-1342-9 (pbk. : alk. paper)

Printed in the United States of America

 ⊖™ The paper used in this publication meets the minimum requirements of
American National Standard for Information Sciences—Permanence of Paper for
Printed Library Materials, ANSI/NISO Z39.48–1992.

For Anne
without whom not

and to the memory of my Mother
Ruth Mitchell Christ
1926 – 2004
Rest in Peace

Contents

Preface   ix

1   Introduction   1

2   Authors and Commons: Two Visions of Intellectual Property Rights   13

3   An Overview of Intellectual Property   25

4   Two Moments in the History of Copyright   33

5   The Natural History of Intellectual Property   45

6   Commons: The Third Form of Property   67

7   The Intellectual Commons   87

8   The Commons in History   109

9   Social Utility and the Rise of the Imperial Author   125

10 The Author Metaphor   139

11 Ethical Issues of Patent Law: Equity and the Intellectual Commons   153

12 Conclusions and Critique   173

Appendix A   Economics: Some Definitions   187

Appendix B   Property, Ownership, and Rights: A Framework   193

Appendix C   The Statute of Anne   201

Bibliography   207

Index   217

About The Author   221

Preface

This book explores intellectual property (IP): the ownership of literary and artistic works, useful ideas, distinctive marks of identity, and commercially useful secrets. To borrow a metaphor from the computer scientist Harlen Mills, the legal theory of intellectual property resembles a cow much more than it resembles a Greek temple: it is the highly functional, absurdly complex, and startlingly contingent product of a long evolutionary process.

This work is an attempt to develop a theory of intellectual property that is based on a theory of natural rights and which assumes the existence of a "natural world" of intellectual resources which I call the intellectual commons. My theory is based on the following principles:

- The objects of IP rights (works) *by nature* cannot be exclusively owned.
- The raw materials of works (ideas) *by nature* can never be owned even though they are made by people through labor and even though they begin in a state of "occupation" (de facto exclusive possession).
- The objects and processes connected to IP *by nature* involve communities and shared resources.
- The objects of IP *by nature* go through a "life cycle" of property relationships. This process creates a domain of renewable intellectual resources: an intellectual commons.
- This intellectual commons is *by nature* infinitely sharable and renewable: only the creation of repressive and perverse property regimes can interfere with this natural intellectual ecology.
- The state should not provide economic incentives to encourage people to do what they are already going to do anyway.
- The state needs to morally justify granting rights that benefit a few greatly by imposing a small cost on many. As with pollution, we can-

not simply externalize the cost because it is diffuse.
- The fact that something *could* have economic value if it were property is not a sufficient reason to *make it* property. Grants of property rights where none existed create social costs that must be morally justified.

From these general principles I derive the following practical conclusions:

- No one has the right to restrict innocent use of the intellectual commons, and analysis of IP rights must evaluate the "environmental impact" of allowing them. This analysis is not purely consequentialist.
- Situations in which the state transforms a commons into private property (enclosure of the commons, homesteading, privatization of socialist economies, expansion of IP rights) must be morally justified in terms of the interests of all parties in the *use* (rather than some static valuation) of the previously common resource.
- The introduction of new types of property (allowing new classes of patents, etc.) require some justification in terms of justice and community, even in the case of new inventions or discoveries. It is definitely not enough to argue that since nothing like X ever existed before, I am not doing anyone harm by granting or claiming property rights in it.

This approach is at variance with the *official* ideology of intellectual property law, which claims that IP systems are purely consequentialist regimes designed to promote a particular kind of socially valuable activity. The official theory is based on the following assumptions:

- Intellectual property rights are a grant from the state designed to encourage authors to share their work with the public. In practice, the more property rights are potentially available, the more authors should be willing to create. In effect, what is good for authors is good for everyone.
- The state needs to provide such an incentive because authorship creates a "public goods problem": works are very expensive to create but very easy to copy and share. Without intellectual property rights, authors would have no incentive to share their works.
- All grants of IP rights must be designed to balance the economic needs of authors with the social value of their works. This balancing is an *economic* problem, not a moral one.

The intellectual commons theory is also at variance with the *actual* ideology of intellectual property law, which Jessica Litman calls "the romantic theory of authorship."

The critical side of the intellectual commons theory defended here is largely inherited from the work of a loose association of legal copyright skeptics, including Jessica Litman, Pamela Samuelson, and James Boyle. The historical basis for their skepticism is made clear in the work of Michel Foucault. The work of the copyright skeptics provides more than enough reasons to question both the official theory and the romantic theory of authorship.

Attempts to build a "post author" theory of intellectual property rights are less common. The most notable are Alfred Yen's "Restoring the Natural Law: Copyright as Labor and Possession" which despite its title attempts to develop a "pluralistic" theory of intellectual property rights.

The intellectual commons theory is similar in approach to the theories mentioned above but different in its philosophical foundations. Hughes and Yen both attempt to build "labor theories" of intellectual property rights: the intellectual commons theory makes labor secondary to a justification based on the natural community of authors and users and the naturally unconfinable nature of ideas. My theory is also pluralistic (i.e., it assumes that there are several different sets of actors, all of whom have unique sets of natural rights).

Perhaps the most distinctive feature of the intellectual commons theory is its historical roots. I was originally led back to the seventeenth century by a desire to understand the historical origins of intellectual property, and to explore the development of the modern institution. Through the Richard Tuck's *Natural Rights Theories*, I found that many of the questions I wanted to explore about concepts of common property were addressed in the writings of Hugo Grotius: as Grotius explicitly recognizes, they go back to Roman law and to its Stoic foundations. Though the intellectual commons theory is based on natural law, it is the non-Aristotelian Roman natural law of Justinian's *Digest*, Cicero's *On Duties*, and Grotius' *On The Freedom of the Seas*.

"Natural law" is perhaps a grander term than what I have in mind here. In the words of Alfred Yen, "there was nothing aspirational about Roman Natural Law. Roman natural law was simply the set of legal rules which corresponded to the simple facts of life. As such, it was a quasi-scientific, inherently rational way of constructing the law." My approach is a "natural law" approach in the sense that I reject legal positivism and I explicitly endorse the idea that laws must have some kind of foundation in moral theory. It is also a "natural" approach because I am assuming that there is a kind of inner order and rationality to the social and legal institutions associated with intellectual property. It is possible to ignore or defy this "nature," but only at the cost of creating institutions that are irrational, unjust or perverse.

Intellectual Property makes the most sense when we see it as moving into and out of a world of its own, which I call the intellectual commons. The intellectual commons contains the raw materials that people use to create works. Like the natural world, it is an arena where people give, take, own, and impose duties on each other. However, the intellectual commons is unlike the natural world in two crucial ways. First, the resources of the intellectual commons are not constrained by discreteness: they can be shared by many people without creating scarcity. Second, the intellectual commons itself has no natural boundaries: it is continually growing through human labor in community. Though the intellectual commons has the given, sustaining character of the physical world, it is entirely the creation of human activity.

The made character of the intellectual commons means that, like the natural world, it can be affected by decisions about sharing and restricting access. Legal regimes that keep it from growing or that create barriers to the use of its resources will have a real impact on its viability. It is an arena where there is a need for a kind of "environmental ethics" to make sure that its resources can be used, preserved, and developed in a just and rational way. As we shall see later, the lawsuit brought by Eric Eldred and others to block implementation of the Copyright Term Extension Act is based in part on a demand for more formal justification of the social costs inherent in strengthening intellectual property rights.

Some theorists of the Internet and the human genome have found themselves drawn to what I will call geographic metaphors. The Internet constitutes "cyberspace," a place where people can meet, work, and exchange resources. John Barlow worries that corporations will colonize it, driving out its original hacker inhabitants. The genome is characterized as a "new frontier," where researchers need to "stake their claims" before everything is taken. Though I will discuss geographic metaphors at various points in what follows, I must confess that I mistrust them. No one can really live in cyberspace: for better or worse, we must always function in what some hackers call RL (real life). And I profoundly distrust frontier metaphors: a frontier is understood by its potential exploiters as *res nullius*, the property of no one. It exists to be dismantled into a world of private property. People don't live in a frontier: they "conquer" it or "push it back." Frontiers are the arena of scarcity and competition, not the arena of any kind of common life.

Intellectual Property Theory has continued to develop as I have prepared this manuscript. I regret not being able to consider several important new works, particularly Jessica Litman's *Digital Copyright*, Lawrence Lessig's *The Future of Ideas*, and Yochai Benkler's groundbreaking work on incorporating commons ideas into the constitutional analysis of IP rights. I also regret the fact that this work has so little to say about trademarks and the right of publicity: the claim that my identity and my privacy should be understood as *property* is at the root of many of the most egregious abuses of IP law. I intend to address these issues more fully in other work.

The intellectual commons theory is pluralistic: authors, users, and publishers all have rights that need to be respected. These rights are unified into a coherent and consistent pattern by the commons concept. The pattern is illustrated here in a fuzzy, low-resolution way: the theory dictates requirements and priorities, not a detailed blueprint. Yet it is distinct enough to mandate changes in the current system. At the very least, legislators and judges need to bring their theory and practice together. Thinking about IP rights also needs a fundamental reorientation away from an obsessive focus on the property rights of authors to a recognition of the reality of user rights and the need to care for our intellectual world as we do our physical world.

A note on the appendices: Appendix A provides an analytical glossary of terms from economics that are used in the text. Appendix B describes a fairly standard definition of property and rights. Appendix C gives the complete text of the Statute of Anne. All of this material is important, but not strictly needed to support the flow of the argument.

Acknowledgments

I would like to thank my wife Anne and our three children, Jesse, Chenoa, and Elise, who helped in uncountable ways and put up with a lot so I could get this project done. I will never be able to repay Anne for supporting our family by working nights so that I could finish my Ph.D. (though I plan to spend the rest of my life trying!). Thanks to all those at Notre Dame who were willing to consider my request to pick this project up again after such a long absence, especially Vaughn McKim for his faith in me and his years of commitment to this project. Thanks to Jim Sterba at Notre Dame and Robert Carley at Lexington Books for encouraging me to prepare this book. An anonymous referee has helped with much constructive criticism. I would also like to thank the staff of the Elkhart Public Library, which has a scholarly depth unusual for a municipal library. My work was also made possible by the generosity of the many scholars who have made their papers available online. On behalf of isolated scholars everywhere I heartily thank you! In many ways the work before you is also a product of the free software movement: I am indebted to the members of that community for providing me with the arsenal of tools I use in my writing, including Linux, KDE, Emacs, TeX, LyX, TEX4HT, and OpenOffice. And a special thank you to James Boyle, whom I hope to meet one day, for getting me interested in the wonderful world of intellectual property.

1

Introduction

As a legal witness, I became conscious of the contradiction between the romantic conception of authorship the notion of the creative individual—that underlies copyright, and the fact that most work in the entertainment industry is corporate rather than individual. Furthermore, many of the characteristic products of the industry game shows, soap operas, situation comedies, police stories, spy stories and the like—tend to be formulaic. Romantic conceptions of authorship seem as inappropriate in discussing these cultural productions as in the equally formulaic productions of some older periods, ballads, say, or chivalric romances. I found these contradictions between the ideology of copyright and the actual circumstances of litigation intriguing and provocative.

—Mark Rose,
Authors and Owners

This work is an attempt to explore the philosophical foundations of the concept of intellectual property (IP). To put it as crudely as possible, IP is a set of social and legal institutions that define property rights in ideas. All societies have institutions that assert control over ideas:[1] examples would include political censorship, obscenity laws, and the esoteric knowledge of religious groups. The defining characteristic of IP is the definition of that control in terms derived from the language of real property ownership. IP theory is thus a branch of the theory of property rights.[2]

IP theory is not unique in the abstract character of its subject matter: contracts and leases are every bit as abstract as copyrights and patents. What does make IP unique as a form of property are the concepts of limitation of term and the public domain. IP is the *only* form of property in which ownership is essentially time-limited. It is also unique in what happens to the property after the term expires: it can no longer be owned by *anyone*. The idea of appropriation, where a thing goes from being unowned to being owned is foundational for many political theories.[3] IP is the only regime in which anything ever goes back to an unowned state.

In what follows, we will use the following basic terminology:

- **Ideas:** The basic raw materials used in the creation of intellectual property.
- **Works:** The individual units of intellectual property. Works will be assumed to have either a physical realization or a type/token relation-

1

ship to a set of physical things or events.

- **Authors:** The persons responsible for creating works. We will use this term for all forms of IP. This person could also be a temporarily and spatially discontinuous collection of persons, or a legal person such as a corporation.
- **Publishers:** The persons responsible for creating physical realizations of the work. These physical realizations can be things or events (e.g., broadcasting a performance). We will use the same broad definition of "person." Publishers are also usually owners of IP.
- **Users:** The consumers of works (though of course that use is not necessarily consumptive).
- **Owners:** Those who have legal control over works. Note that ownership of IP is fully alienable, so the owners are often publishers rather than authors.

Note that there is no particular reason to assume that the last four categories are disjoint: authors can publish their own works (as Daniel Defoe did), and everyone is a user at one time or another. It is also possible for publishers not to be owners (U.S. government publications are not copyrighted).

Philosophy and Intellectual Property

Jessica Litman and James Boyle argue that the expansion of IP rights is a product not only of new economic interests but also of a one-sided, "author-centered" understanding of IP theory that ignores an author's debt to previous authors for her starting material.[4] Their critique is based in part on a revisionist historiography of intellectual property developed by Bernard Kaplan, L. Ray Patterson, Mark Rose, and Michel Foucault.[5]

Philosophers have had relatively little to say about intellectual property, despite the fascinating puzzles it engenders. Becker's *Property: Philosophical Foundations* contains only a single mention,[6] while Waldron's *The Right of Private Property* uses intellectual property as an example of intangible property without analyzing it further.[7] I know of only two survey articles on philosophical aspects of IP.[8] Papers about IP have generally been critical: some have argued that the concept of copying itself is incoherent,[9] while others have argued that the existing IP framework has unjust consequences, particularly for people in the developing world.[10] On the other hand, philosophers throughout history, particularly the seventeenth- and eighteenth-century natural-law theorists, have had a great deal to say about appropriation as a source of property rights.[11] The time is ripe for the development of a non-author-centered theory of intellectual property. Such a theory must be based on understanding what is now called "the Public Domain," and would shift the focus of authorship from creation *ex nihilo* to the appropriation of material from an intellectual commons. Such a theory would focus not only on authors' rights, but also on authors' obligations to protect and expand the intellectual commons. The goal of this work is to develop such a commons-based theory of intellectual property.

Though the problem of IP is particularly acute in our time, I believe that the conceptual tools we need to develop a new theory were actually developed long ago, in the time of Grotius, Pufendorf, and John Locke. These philosophers struggled with creating truly radical theories of property: theories that could ground property rights in a general moral theory instead of simply taking their nature (and distribution) for granted. In order to develop a new approach, they told stories about beginnings (the state of nature). These stories focus on a moment when things began to move from being not-property to property. A deeper understanding of such moments is the key to understanding intellectual property.

Ethical Constraints on Intellectual Property Systems

Intellectual property thus involves a unique relationship between the owners of IP and the state. Despite the fact that IP is unquestionably a product of human labor and creativity, it can only be owned for a fixed period of time. Even during the time of ownership, the state protects the owner's title but not does not grant her exclusive control: others may still use the work in various acceptable ways. What could possibly justify such an unusual combination of public and private rights? A "concept" of IP is justified in terms of some more basic theory of property. A "conception" of IP (e.g., copyright) must be justified in terms of the concept and must also dictate specific practices, which we will call "implementations." As Rawls puts it,

> Men disagree about which principles should define the basic terms of their association. Yet we may still say, despite this disagreement, that they each have a conception of justice. That is, they understand the need for, and are prepared to affirm, a *basic set of principles for assigning rights and duties and for determining what they take to be the proper distribution of the benefits and burdens of social cooperation.* Thus it seems natural to think of the concept of justice as distinct from the various conceptions of justice and as being specified by *the role which these different sets of principles, these different conceptions, have in common.*[12]

We will generally use the term regime for the combination of a conception and its implementation.

Our theory will be based on two claims and two constraints:

- It is meaningful to allow property rights in works and ideas (possibility).
- It is possible to construct a system of IP rights that is not morally arbitrary (legitimacy).
- The legal and institutional framework designed to implement IP rights must not be self-defeating: that is, consistent practice of the implemen-

tation should not undermine the values it is intended to embody (coherence constraint).
* Property rights in ideas must have limited scope and limited term (limitation constraint).

Asserting the possibility condition involves going beyond what might be seen as
natural or primitive concepts of property,[13] since there can be no occupation or
physical control of an idea. The theory of real property has the advantage that its
object can be enclosed in some sort of objectively verifiable way.[14] In the words
of David Lange, "The chief attribute of intellectual property is that apart from its
recognition in law it has no existence of its own."[15]

It is important not to exaggerate the differences between real property and IP
here. All property rights must ultimately be backed by the coercive power of the
state. We must also recognize that the possibility of a system of IP does not require that all questions about identity and individuation be answered in advance:
the law often proceeds by positing a vague general right that is iteratively sharpened by the opinions of judges. Even more: IP systems are generally built on the
assumption that not all boundary issues can be settled in advance, but must be
dealt with on a case-by-case basis.

The legitimacy condition is important because all forms of IP have been challenged on moral grounds, and often for good reason. The first English patent system degenerated from a form of industrial policy to a corrupt system of royal patronage, with entrepreneurs being granted patents on an entire, pre-existing
industry (e.g., patents for making glass or soap).[16] These patents were simply
sold to the highest bidder, whether or not they had actually advanced industry in
any way or not. In pre-revolutionary France, copyrights could be awarded as a
form of patronage without involving the actual author at all.[17]

IP can be morally problematic because it functions as a monopoly backed by
the coercive power of the state; in effect, the state is awarding rights to some at
the expense of all others. In a society committed to *de jure* equality, awarding
such rights requires justification.

The coherence constraint is a constraint to the implementations of IP systems,
but certainly not something restricted to intellectual property. The demand for
coherence is the demand that we recognize the contradictory forces often created
by systems of law and property. Antitrust regulations constrain the free operation
of markets so that the preconditions of such a market will continue to exist. Only
the most utopian continue to believe that market economics or central planning
will automatically produce an optimal economy. Note that the coherence constraint can be read in either a consequentialist or deontological sense. We might
want implementations to promote or honor the values that justify them, or both.

Two conceptual threads run through the argument that follows. The first is
that all discussions of intellectual property involve distributive justice. IP rights
are economic monopolies backed by the coercive power of the state: the author's
gain is the user's loss. Whether we justify IP rights as tools of economic policy

or as the state protecting the natural property rights of authors, it is undeniable that IP rights limit the options of users and other authors.

IP can never be mine in the sense that the shirt on my back or the change in my pocket is mine. There is no intellectual property in the state of nature, even if there is property in acorns and apples. The rights, obligations, and sanctions of an IP regime only make sense in the context of a broader political system. This regime cannot simply be modeled on the ownership of tangible things, because the exclusivity inherent in the possession of tangible things does not apply to intellectual property. Decisions about how much an inventor or artist gets to "hold onto" her work (through either economic rights or rights to control use) cannot be answered by a simple causal account like Nozick's entitlement theory.[18] The dimension of distributive justice gets lost when we focus on IP either as a natural right of authors or as a legal fiction that is purely instrumental in character.

In this work we will call a system of IP rights monolithic if it makes the assumption that all IP rights are derived from the rights of one group. A theory is pluralistic if it is based on granting rights to several or all of the groups involved in an IP system.[19] Using the distinctions made above, there seem to be at least five possible theories of IP rights.

- **Utilitarian/Instrumentalist:** An IP system provides incentives for authors to create works of general utility and value by giving them property rights in what they create. On this view, the property rights of authors are a convenient fiction that is justified only by the social utility of the whole system. It would be possible to interpret this approach as an antitheory: the claim that IP systems have no intrinsic philosophical interest, since they are simply contingent procedural arrangements.
- **Author-centered:** The IP system exists to protect the rights of authors. These rights could be justified on either a consequentialist or natural rights framework.
- **Publisher-centered:** The IP system exists to make it possible for entrepenuers to invest in the production of IP and have some assurance that their investment will not be stolen or expropriated.
- **User-centered:** The IP system exists to promote the greatest possible access to works by users.
- **Pluralistic Theory:** The correct theory is some combination of the above: authors, users, and publishers all have rights that must be simultaneously accommodated. IP must be worked out in an environment of balancing interests rather than zero-sum conflict. The resulting system will never be ideal.

The historical development of intellectual property law has been driven by changes in technology to a much greater extent than the development of other forms of property law. New technology creates new media for the dissemination of works (e.g., player piano rolls, radio, television, the World Wide Web) and powerful methods of copying (e.g., Xerox machines, VCRs, bulk CD copying machines). The difficult issues raised by media and copying technology can be seen as evolutionary extensions of the problems originally posed by print publishing.

Other changes are clearly revolutionary. The end of the twentieth century has seen the emergence of totally new forms of intellectual property. Forms of life and algorithms can now be patented. Farmers who save some of their genetically engineered crop to replant are guilty of patent infringement. Physicians are beginning to patent medical procedures as well as medical devices.[20] Copyright protection has been extended far beyond creative works: "Copyright vests automatically in your shopping lists, your vacation snapshots, your home movies, and your telephone message lists."[21] Distinctive features of a person's appearance and behavior (e.g., Bela Lugosi's cape, Charlie Chaplin's cane and waddle) are now considered intellectual property.[22]

Proponents of new IP rights typically argue that these changes are simply extensions of traditional rights into new areas. The authors of the *National Information Infrastructure (NII) White Paper on Intellectual Property* explicitly claim to be doing nothing new: "Existing copyright law needs only the fine tuning that technological advances necessitate, in order to maintain the balance of the law in the face of onrushing technology."[23] Their critics see vast new property claims in this "fine tuning." [24]

Critics of new IP rights typically argue that these new grants tend to undermine the IP social contract by shifting more and more power away from users and toward authors.

Authors, Publishers, and Users

Copyright has been traditionally understood in terms of a triadic relationship between authors, users, and publishers (the middlemen who mass-produced the author's work). All three corners have "evil twins": plagiarists, free riders, and pirates. Free riders and pirates infringe the economic rights of authors (and publishers); plagiarists infringe the moral rights of authors.

The economic interests of publishers and authors can coincide: each depends symbiotically on the other. However, huge power imbalances are possible. The Stationer's Company didn't pay royalties, only a flat (usually small) fee.[25] In mass media, stars can create their own power imbalances: extremely popular ones can demand enormous (even ruinous) fees for their work.

The relationship between users and authors/publishers is significantly more ambivalent. Each needs the other, but they are locked in an economic zero-sum game. Users want to get as much access as possible for the lowest cost; producers want to maximize their economic gain. Authors may or may not benefit from the publisher's gains.

Both authors and users sometimes perceive the relationship with publishers to be an onerous necessity and a historical accident. One of the most radical possibilities created by digital media is the possibility of eliminating publishers and allowing a direct connection between authors and users. However, we should recognize that elimination of publishers would also eliminate the social controls built into the current system: publishers are censors either directly or indirectly. The economic and social interests of publishers makes them useful instruments of political/social policy.

It should also be recognized that the emergence of digital media could also be used to greatly expand the role of publishers. Some have argued that computer technology could be used to give publishers control over every use of every copy of a work. The required technology (trusted systems) is currently being tested.[26] Several commentators have decried this power shift between users and publishers.[27]

The Essence of Intellectual Property: Limitation of Term

All theories of property must address the problem of appropriation: how is it that unowned things become property? IP theory is unique in insisting we must also consider "antiappropriation": the movement of owned things into the status of unowned things. IP theory actually is far more radical than this: the newly unowned things become unownable. They seem to constitute some kind of "antiproperty": a sort of intellectual commons.

Limitation of term is a complicated idea that invites at least three possible interpretations. The first is frankly skeptical about the use of the property model: limitation of term is the recognition that IP is not really property at all, but simply a state-sanctioned monopoly that is justified by its contribution to the social good. Revocation of such a right is not a taking (e.g., an injury that requires compensation), but simply a matter of balancing the social costs and benefits of the IP regime.

At the opposite extreme would be private property theorists who would argue that limitation of term is analogous to eminent domain: it represents society depriving an individual of their property in order to realize some social good, and thus requires compensation of the author. IP policymaking thus becomes a process of individuals (authors) resisting government encroachment on private property. This is the position of the *NII White Paper*, where changes in existing IP rights are rejected because they are seen as unfair subsidies to the social good at the expense of authors.[28]

Both extremes agree that IP policy realizes some social good. But what is that good? The Constitution specifies it as "The advancement of science [that is, art and science] and the useful arts [technology]." Why does the advancement of science require limitation of term? Because authors do not create ex nihilo, but build on the foundation provided by earlier authors. Without limitation of term, IP would become an oligarchy dominated by those who got there first. The haves of the IP system would be in a position to control or block the creative efforts of the "have nots" by demanding compensation.

The defenders of IP rights argue on consequentialist grounds when pressed to justify the institution of intellectual property. According to the classical argument, authors are more likely to create and share their works if they are guaranteed some form of exclusive ownership for their works.

The perspective laid out above provides a natural explanation for the importance of limitation of term. If IP rights were open-ended, the requirement that authors share their works would only be partially met. Unless future authors can use existing works as raw material, the process of authoring will become more and more constrained by the IP rights of the past. Limitation of term gives authors the economic rights they need, while guaranteeing that their works will not ultimately constrain the work of future authors.

A really whole-hearted embrace of this consequentialist argument could be seen as providing a deeper explanation of limitation of term. If the ultimate justification of IP is the progress of science and industry, then IP isn't really property at all: it is a social gratuity that is justified by its social consequences. One might expect that a consequentialist would be inclined to support such a line of argument.

The Instrumentalist Objection

Consider the following objection to a philosophic theory of intellectual property: "IP is purely a pragmatic issue of public policy. Why should IP law be more philosophically interesting than, say, traffic laws? No one would dispute the utility of having speed limits or having everyone driving in a particular direction drive on the same side of the road, but there is unlikely to be much philosophical meat in debates about whether that speed should be fifty-five or sixty-five miles per hour. Why should there be anything philosophically interesting about the fact that the term of copyright was raised in 1998 from life of the author plus fifty years to life of the author plus seventy years?" We will call this position instrumentalism: the instrumentalist argues that IP theorizing is uninteresting since it is merely an argument about tactics.

The basic response to the instrumentalist objection is a general point about instrumentalism itself. An instrumental account argues that the sole justification for some institution is its promotion of some outcome: all questions of detail are merely tactical. Stated in these terms, it becomes obvious that *global* justification of an institution on purely instrumental terms is impossible. An instrumental justification argues that an institution is justified because of the way that it promotes some particular state of affairs. But why should we promote *this* state of affairs? There must be some higher-order reason to prefer it over others, and it is hard to see how this higher-order justification could avoid the language of values in favor of the language of tactics.[29]

The instrumentalist objection seems to be based on two mistaken ideas about political philosophy. The first is the idea that analysis of a conception is the same as analysis of an implementation. But even a little reflection makes it clear that a conception can have many possible implementations. Even the effort to analyze a regime in "purely pragmatic" terms must appeal to some other set of principles to justify *this* choice of implementation.

The second mistaken idea is a bit more subtle. The instrumentalist seems to be arguing that IP theory is inherently uninteresting because it can be (or must be) analyzed purely as a matter of tactics. This refusal to leave the level of the practical is most charitably explained as the outgrowth of the belief that conceptions "emerge" from concepts and thus cannot be reductively analyzed. But this belief confuses *explanation* with *justification*. It certainly seems reasonable to reject a "top-down" theory of IP which assumes that conceptions are totally determined by concepts. But this is a different matter entirely than the effort to justify the legitimacy of a conception in terms of a concept. My claim is that the concept/conception distinction is reasonable, and that no political theory is complete until conceptions and implementations are justified both upwardly (in terms of concepts of justice) and downwardly (in terms of economics and moral psychology).

The Classic Consequentialist Argument (Incentive Theory)

The practical justification for IP rights has been stated very clearly by Jessica Litman:

> According to a currently popular mode of analysis, property rights in intellectual works are necessary because intellectual creations pose a public goods problem: The cost of creating works is often high, the cost of reproducing them is low, and once created, the works may be reproduced rapaciously without depleting the original. In a world in which such reproduction is not restrained, an author will be unable to recover the costs of creating the work and will therefore forgo the creative endeavor in favor of something more remunerative. To provide the author with a market in which she can seek compensation for her creation, we establish property rights in her work and allow her to sell or lease these rights to others. Thus, the copyright system encourages authors to create and encourages distributors to purchase rights in author's creations so that the distributors may sell these creations to the rest of us. [30]

As we shall see, this rationale is not just "currently popular." Moreover, for our purposes it is important but inadequate, since it is a pragmatic justification that presupposes the moral value of innovation.

There seem to be two classic approaches to morally justifying IP rights. The first approach is in terms of a social contract ethic. It is in the interest of society as a whole that authors continue to create new works and inventions. To facilitate this common good, the state creates intellectual property rights so that authors can get a return for their efforts. In return, authors publicly disclose their work so it can be used as a basis for other new works. Since granting authors absolute and perpetual property rights would undermine the goal of public use, the property rights are only granted for some fixed period of time. Thus, the U.S. Constitution grants Congress the power to "*Promote the Progress of Science and useful arts*, by securing for limited Times to Authors and Inventors the exclusive Right to their respective Writings and Discoveries."[31] IP rights represent an implicit social contract between authors and users.

We should note that the social good of authorship (the progress of science and industry) functions as a final cause for the existence of IP rights, but the efficient cause for the existence of these rights is the need to correct the "market failure" associated with the nature of works as public goods. This has tempted many to argue that the justification of IP rights is purely economic.

Natural Rights Theories

There is a second, less familiar justification for IP rights that I call a "natural rights" theory of IP. A theory of this type posits (or blocks) IP claims because of perceived connections between IP and some broader moral context (e.g., the nature of creative activity or the nature of communities).

There are two levels of distinction that are important here. At the broadest level, natural rights theories represent an alternative to social contract theories. The two theories differ most clearly on the question of the origin of IP rights. In the natural rights theory, the state recognizes IP rights; in the social contract theory, the state creates them. There is also a significant practical difference: in the natural law theory, IP rights should be permanent in the same way as other property rights. The social contract theory (and virtually all IP laws) sets time limits on IP rights, after which the works are available to anyone.

There is another distinction that is more important for our purposes. There are two extant natural rights theories of IP rights. The most familiar is the foundation of author-centered theories. According to this view, the works created by an author should be his property (in the sense of full liberal ownership) because he is their creator. In this view, there is (or should be) no significant difference between owning a copyright and owning a piece of furniture: it is attempts to circumscribe such rights that requires justification, not the rights themselves. IP rights are simply a form of individual property rights.

Natural rights have been a muted but constant undercurrent in the development of IP theories, a process which has been dominated from its beginning by social contract theories. We will examine some consequences of a natural rights approach in chapter 3, particularly the prominence it gives to originality and creativity.

Notes

1. Social control of Ideas is discussed in Michel Foucault, *The Order of Things* (New York: Random House, 1970); Tom Greaves, *Intellectual Property Rights for Indigenous Peoples* (Oklahoma City, OK: Society for Applied Anthropology, 1994); and Roger Shattuck, *Forbidden Knowledge: From Prometheus to Pornography* (New York: St. Martin's Press, 1996).

2. This categorization allows overlap with other regimes of information control, as we shall see when we discuss privacy as property. It also accepts that IP laws can be used as a means to some other form of intellectual control, as when the Church of Scientology uses copyright laws to suppress publication of their religious secrets by former members.

3. The best-known Lockean theory of property rights is that of Robert Nozick, *Anarchy, State and Utopia* (New York: Basic Books, 1974). James Tully, *A Discourse on Property: John Locke and His Adversaries* (New York: Cambridge University Press, 1980) attempts to move Locke's theory beyond a simple identification with possessive individualism.

4. Jessica Litman, "The Public Domain," *Emory Law Journal* 39 (1990): 965. See also James Boyle, *Shamans, Software and Spleens* (Cambridge, MA: Harvard University Press, 1996).

5. Benjamin Kaplan, *An Unhurried View of Copyright* (New York: Columbia UniversityPress, 1967); L. Ray Patterson, *Copyright in Historical Perspective* (Nashville TN: Vanderbilt University, 1968); Michel Foucault, "What is an Author?" in *The Foucault Reader*, edited by Paul Rabinow (Laguna Niguel CA: Pantheon, 1969); Mark Rose, *Authors and Owners* (Cambridge MA: Harvard University Press, 1993).

6. Lawrence Becker, *Property Rights: Philosophical Foundations* (New York: Routledge, 1977).

7. Jeremy Waldron, *The Right to Private Property* (New York: Oxford University Press, 1988).

8. Edwin C. Hettinger "Justifying Intellectual Property," *Philosophy and Public Affairs* 18: 31–52. See also Mark Alfino "Intellectual Property and Copyright Ethics." *Business and Professional Ethics Journal* 10 (1991), no. 2: 85–109.

9. Selmer Bringsjord, "In Defense of Copying" *Public Affairs Quarterly* 3 (1989):1-9.

10. Brian Martin, "Against Intellectual Property," *Philosophy and Social Action* 21 (1995) vol. 3: 7–22. R. S. Glass and W. A. Wood, "Situational Determinants of Software Piracy: An Equity Theory Perspective," *Journal of Business Ethics* 15, (1996): 1189–1198, and W. R. Swinyard, H. Rinne, and A. K. Kau, "The Morality of Software Piracy: A Cross-Cultural analysis," *Journal of Business Ethics* 9 (1990): 655–664.

11. The definitive study of early modern natural law is Richard Tuck, *Natural Rights Theories: Their Origin and Development* (New York: Cambridge University Press, 1978). See also Stephen Buckle, *Natural Law and the Theory of Property: Grotius to Hume* (New York: Cambridge University Press, 1991).

12. John Rawls, *A Theory of Justice*, Revised Edition (Cambridge, MA: Harvard Belknap, 1999), 5, emphasis added. Rawls himself attributes this distinction to H. L. A. Hart.

13. Becker, *Property Rights,* 24–30.

14. Charles Donahue, "Property Law." *Encyclopedia Britannica* (15th ed.), Volume

26, 180–205. (Chicago, IL: Encyclopedia Britannica, 1997).

15. David Lange, "Recognizing the Public Domain," *Law and Comtemporary Problems* 44 (1981) no. 4: 147–178.

16. W. H. Price, *The English Patents of Monopoly*, Volume 1 of *Harvard Economic Studies*, (Cambridge, MA: Harvard University Press, 1913).

17. Jane Ginsburg "A Tale of Two Copyrights: Literary Property in Revolutionary France and America," in Brad Sherman and Alain Strowel, *Of Authors and Origins: Essays on Copyright Law* (New York: Oxford University Press, 1994), Chapter 7, 131–158.

18. Nozick, *Anarchy, State, and Utopia,* 150–153

19. See L. Ray Patterson and Stanley Lindberg, *Copyright: A Law of User Rights* (Athens, GA: University of Georgia Press, 1991) for an example of a pluralistic theory.

20. Seth Shulman, "Cashing in on Medical Patents," *MIT Technology Review* 101 (1998) no. 2: 38–45.

21. Jessica Litman, "The Public Domain," 974.

22. David Lange "Recognizing the Public Domain," 148.

23. Bruce Lehman et. al., *National Information Infrastructure Task Force Working Group on Intellectual Property Rights Final Report* (known as "The NII White Paper." U.S. Patent and Trademark Office, September 1995), 17.

24. John Boyle and Bruce Lehman, "Debate on the NII White Paper" (correspondence between Boyle and Lehman, posted on the internet by Boyle). http://www.w-cl.american.edu/pub/faculty/boyle/boyledeb.htm, January 21, 1998.

25. John Milton was paid a total of about five pounds for *Paradise Lost* by his publisher. Rose, *Authors and Owners,* 27–28.

26. See Mark Stefik, "Shifting the Possible: How Trusted Systems and Digital Property Rights Challenge us to Rethink Digital Publishing," *Berkeley Technology Law Journal* 12 (1997) no. 1.

27. Jessica Litman, "The Exclusive Right to Read," *Cardozo Arts and Entertainment Law Journal* 13 (1994): 29.

28. *NII White Paper,* 84 (note 266).

29. This is, of course, a familiar argument from Aristotle: see *Nichomachean Ethics,* 1094a,15–25.

30. Jessica Litman, "The Public Domain," 970.

31. U.S. Constitution, Article I, Section 8, Clause 8. Emphasis added.

2

Authors and Commons: Two Visions of Intellectual Property Rights

The dialectic of intellectual property rights is driven by the interaction of three conceptions: a pragmatic or economic point of view, a view that focuses on the property rights of creators, and a view that focuses on the uncircumscribed nature of ideas and the inherently communal nature of the creative process. The first point of view is the typical ideology of legislators, the second that of authors and publishers, and the third that of "users."

The dialectic is subtle because many actors espouse one point of view while acting in accordance with another. The stated rationale of new IP rights is generally presented as a pragmatic bargain: IP grants are presented as a state correction to "market failures" or "public goods problems" in the interest of encouraging the dissemination of useful and valuable works. If IP rights were truly pragmatic, one would expect at least some of the adjustments to favor users instead of owners, or at least an attempt by the legislators to empirically measure whether or not existing IP regimes are doing what they were supposed to do. Yet adjustment of existing IP rights almost always goes in the direction of making them more valuable to their holders, and the kind of cost-benefit analysis that dominates other forms of regulation is virtually nonexistent.

This outcome reflects more than the raw political power of publishers. It also reflects the extent to which a debate about the exercise of state power (the creation of a legally enforced monopoly) has been recast as a debate about defending individual property rights. While some commentators are inclined to see this recasting as a cynical ploy,[1] our view is that its success reflects genuine moral insight. There is certainly some sense in which a work "belongs" to an author, some sense in which that "ownership" is morally legitimate. It will be the burden of this book to argue that there is another, equally valid set of moral rights implicit in the IP system: the rights of users. Honoring these rights is a matter of morality, not pragmatics.

Debate about IP law tends to be dominated by legislators and publishers, since they form a well-defined constituency with clearly delineated interests. Except for some specialized institutions (most notably universities and libraries), users have lacked any unified voice. As a result, IP law almost exclusively favors publishers and authors. The rights and needs of users, if they are recognized at all, are given cursory acknowledgment and then ignored. Statutes and judicial rulings explicitly address the rights of owners, but the very existence of user's rights is controversial: "as a technical matter, users are not granted affirmative 'rights' under the Copyright Act; rather, copyright owner's rights are limited by exempting certain uses from liability."[2]

Those who wish to attack author-centered theories of IP have several options. The first is to reject the legitimacy of IP systems entirely. Few have gone this far (though Brian Martin and Vandana Shiva come close). We will not take this approach because we accept the moral legitimacy of at least some IP rights. A second critical stance is to promote a different kind of monolithic theory: IP should be user-centered instead of author-centered. The problem with this approach is that, like the author-centered theory, it assumes that IP rights are a zero-sum game: authors get their rights by taking rights from users and *vice versa*. Embracing zero-sum logic fuels the belief that the correct way to defend the claims of [users/authors] is to attack the moral claims of [authors/users]. Defenders of the status quo point to the greed and irresponsibility of some users, while critics of the status quo point to the greed and repressiveness of some owners. Both sides have a point, but it is only a starting point. There will be no progress in debates about IP until each side can defend its own approach in positive terms. All parties concerned with IP can agree on at least this point: works are good, and the creation and dissemination of works is a socially useful thing to encourage.

A great deal of time and energy has gone into promoting the rights of authors and publishers, and we shall be examining their theories. Much less has gone into building a positive theory of users' rights (a defect we hope to partially remedy). But all theories are built around central metaphors and images. We will begin by examining three powerful metaphors from the history of intellectual property.

Goldstein: Making a World Out of Thin Air

> Copyright, in a word, is about authorship. Copyright is about sustaining the conditions of creativity that enable an individual to craft out of thin air, and intense, devouring, labor, an *Appalachian Spring*, a *Sun Also Rises*, a *Citizen Kane*.[3]

It would be possible to create a similar account based on the unique activities of the author; to explain the significance of a work "by discerning, in the individual, a 'deep' motive, a 'creative' power, or a 'design,' the milieu in which writing originates."[4]

Either account makes the author totally central to understanding works. Following Jessica Litman, we will call any theory of authorship grounded in the personality and unique creative acts of an author a *romantic* theory of authorship.

"According to the romantic model, creative processes are magical and are, therefore, likely to produce unique expression. The expression is unique because the real author is using words, musical notes, shapes or colors to clothe impulses that come from within her singular inner being."[5]

A romantic theory of authorship makes originality the most important characteristic of a literary work. Romantic models of authorship interpret the process of creating works in two ways. The most radical is to argue that the author creates *ex nihilo*. A more modest approach is to describe creation as a process of taking pre-existing materials and "stamping" them with the unique perspective of the writer.

The romantic model, by insinuating a kind of organic connection between an author and work, creates (or perhaps justifies) an impetus toward some kind of strong property rights. It seems a short step from saying this painting is "mine" (an expression of my artistic genius) to saying it must be mine (my exclusive property).

Goldstein does not see authorship as simply the passive production of the fruits of genius: in addition to "thin air," the process requires "intense, devouring labor." This labor further grounds the author's ownership of her works on the familiar Lockean account of appropriation. This labor gives the author moral claims against the "free riders" who want to use the work without compensation. Goldstein's choice of adjectives is seems designed to bring out the fact that the creative process is work: something that no economically rational agent would undertake without compensation. It is "devouring" because the work of creation preempts other more routine forms of economic activity. In short, the author, like the entrepeneur, sacrifices present security and satisfaction in the hope of some future gain. The Italian Giacopo Acontio made the following argument to the court of Elizabeth I in 1559: "nothing is more honest than that those who by searching have found out things useful to the public should have some fruit of their rights and labors, as meanwhile they abandon all other means of gain, are at much expense in experiments, and often sustain much loss."[6]

If Goldstein's picture is accurate, it seems plausible to argue that IP regimes must be based on the natural rights of authors (in the minimal sense of "natural law" described in the preface). The natural rights of authors could be expressed in a number of ways, including the following:

- Authors are the conceptual and moral center of intellectual property theory. All moral imperatives associated with IP must be derived from the moral rights of authors.
- Authors have a natural right to property in their works. They have no more obligation to share their intellectual property than they do to share their tangible property.
- Authors have no moral debts in their authorship because they create works "out of thin air."

If IP law is only about authorship then the task of crafting IP law is simply the the task of "sustaining the conditions of creativity" by allowing authors to direct their energies toward the creation of works rather than earning a living. In the next chapter we will consider some of these arrangements: patronage, prizes for useful inventions, honors and status, government jobs and subsidies. All such arrangements hinge on a recognition of an author's right to acknowledgment (so-called "moral rights of authorship"). The most general solution is to create conditions in which authors have an economic interest in the dissemination of their works (so-called "economic rights").

Authors have certain rhetorical advantages in debates about IP rights. They combine both heroism and vulnerability: a lifetime's tireless labor can be stolen in an instant. And no one can deny that authorship is a necessary condition for the existence of works. At the same time, an exclusive focus on the author as the source of IP rights would be misleading. The vast majority of wealth created by authorship flows to publishers, not authors. Most authors are either employees who have no property rights in their work or independent contractors who sell their rights to their employers or publishers. As Mark Rose points out at the beginning of chapter 1, many works are created corporately, with no single author.[7]

There is one point at which the romantic theory of authorship is simply false: the claim that authors create their works "out of thin air." As we shall see, authors rarely if every create works of complete originality, a point that leads to our next picture.

Jefferson: My Taper Burns in the Unconfinable Air

If nature has made any one thing less susceptible than all others of exclusive property, it is the action of a thinking power called an idea, which an individual may exclusively possess as long as he keeps it to himself; but the moment it is divulged, it forces itself into the possession of everyone, and the receiver cannot dispossess himself of it. Its peculiar character too, is that no one possesses the less, because every other possesses the whole of it. He who receives an idea from me, receives instruction without lessening mine; as he who lights his taper at mine, receives light without darkening me.That ideas should freely spread from one to another over the globe, for the moral and mutual instruction of man, and improvement of his condition, seems to have been peculiarly and benevolently designed by nature, when she made them, like fire, expansible over all space, without lessening their density at any point, and like the air in which we breathe, move, and have our physical being, incapable of confinement or exclusive appropriation. Inventions then cannot, in nature, be a subject of property. [8]

In this remarkable passage, Jefferson is making both a narrow, practical point and a broader moral point. The practical point is that there is no way to naturally "fence in" a publicly known idea, since it is available to anyone who understands it: ideas are "incapable of confinement."

Jefferson's negative point (ideas cannot be confined) is matched by a positive one (ideas are infinitely reproducible without loss). As he points out, if someone else lights their taper from mine, mine is still lit. The fact that someone else uses, enjoys, or appreciates my works does not in any way prevent me from using, enjoying, or appreciating them. His broader moral point is that the immunity of ideas from confinement is a good thing that reflects the benevolent design of the world, since "no one possesses the less, because every other possesses the whole of it." *IP is the only domain of property theory where it even possible to imagine a regime without scarcity.* We will call this domain the intellectual commons.

However, it is also possible to use natural rights arguments to argue against the claim that IP rights are individual property rights. As Wendy Gordon puts it,

> [a natural rights approach] suggests that the general population is entitled to demand that the intellectual property law make them whole when it makes incursions into those of the publics "compensable" interests that are protected by natural rights. . . .although natural rights give some support to what proponents of expansive intellectual property call "author's rights," they also give support to the general population and the population of creative users who need to employ other's work.[9]

Morality and Community

The familiarity of "individual rights" arguments creates a strong pressure to use rights as our starting point. Yet we feel uncomfortable with a purely rights-based approach. Most of this discomfort arises from a conviction that the only really useful way to understand IP involves the notion of a community.

Formulation of IP theory solely in terms of individual rights invites conflict into the very heart of the theory. If a right is some kind of indivisible power of a sovereign individual, then rights are zero-sum: I can only have more if you have less. Either I am the embattled author, defending my claims from a hostile and rapacious world, or I am a user, oppressed and deprived by the monopoly of authors. Perhaps the best that can be hoped for is a local optimum that minimizes the frustration of each party. This is the kind of order that is endured, not embraced.

Our deepest instincts tell us that all the parties involved in the world of IP need each other. This sense of need is not merely instrumental: it is somehow essential to the conception of works and their careers in the world. Authors cannot create without participating in a community of practice and a pre-existing world of ideas. Authors have no reason to create without some belief that what they create will somehow be significant or useful to others. The rest of us have to face the reality that works and ideas don't fall from the sky like manna: they are made through the work and sacrifice of dedicated people. This circle is not a lucky co-incidence: it is an organic reality that is a key part of the idea of community.

We must resist the temptation to understand communities as super-individuals who simply replicate the individualistic struggle at a higher level of abstraction. If we succumb to this temptation we will find ourselves again trapped in the mire of zero-sum arguments. We will find ourselves fighting for or against the rights of "the users" or "the authors" for some indivisible good. Perhaps the greatest irony of such an approach is that it gives us the worst of both worlds: we inherit the problems of individualism while necessarily ignoring the reality of individuals. In his critique of utilitarianism, Rawls argues that the greatest flaw of calculations of utility is the fact that such calculations necessarily ignore one fundamental reality: "The correct decision is essentially a matter of administration. This view of social cooperation is the consequence of extending to society the principle of choice for one man, and then, to make this extension work, conflating all persons into one through the imaginative acts of the impartial sympathetic spectator. Utilitarianism does not take seriously the distinction between persons."

Saying that we need to avoid the trap of individualism (manifest or covert) is one thing: doing it is quite another. Perhaps we can begin by a different characterization of boundaries. The natural boundaries of skin and kin divide the world into "mine" and "yours" (or "ours" and "theirs"). But this distinction itself presupposes another more basic reality of "being-in-the-world": the distinction between "someone's" and "no one's." In an individualistic view of the world, this "no one's" is a surd, a purely negative concept. If it is noticed at all, it is as a potential source of future conflict between rival claimants. Its only fate is to be devoured by the "real world" of yours and mine. The individualistic understanding is grounded in the scarcity of resources and the necessary indivisibility of many physical things. But another understanding is possible.

Public Goods

Human life would be impossible in a world where everything was private property (where would the poor stand?). Even in a world dominated by the ethos of private property, there must be public goods: goods that are used by individuals but cannot be paid for individually. The paradigmatic public good is a lighthouse: it costs something to build and maintain, and may override some individual property rights. Yet without it, the entire web of relationships that upholds commerce and private property would be impossible. For economists "public good" is a term of art, describing a lacuna in free-market systems: but we would be well advised to listen to the plain, literal resonance of the term. There is a third realm between the "ours" and "no one's": the domain of the public. And this third domain is good in at least two senses; it is valuable and necessary, but we maintain the term also reflects a kind of moral urgency. To live, we must be parts of communities, and communities cannot exist without the willingness to (at least occasionally) act without calculating individual advantage. We live in a society that is slowly being destroyed by the belief that sharing is an onerous and entirely optional burden.

The usual solution to the dilemma of "cost without fees" is for such public goods to be built and maintained by the state, and paid for through taxation. Even libertarians support the existence of a "night watchman" state that provides basic security and basic infrastructure like roads. There is a wide consensus that the state should maintain free public education and perhaps avert what Nozick calls "moral catastrophes" like famines and plagues. It is certainly possible to imagine a much wider expanse of public goods provided by the state and paid for through taxation.

Appeal to the state might appear to be a simple answer to the problem of public goods, but there are reasons to believe that it cannot be the whole answer. To focus attention solely on the state is to attempt to "reify away" community by identifying it with a concrete institution. Such an identification can clear the way for continuing the same individualistic framework of analysis, now with "community" being a super-person named "state." Whatever must be shared is property of the state (or the state at least acts as trustee of it). A romantic might see the state as a disinterested referee, mediating the conflicting claims of individuals. A cynic might see the state as a powerful bully whose mediation takes place in the context of protecting the state's own interests. Regardless of attitude, the state solves common goods problems by taking from some and giving to others. We "help" some by harming others.

This kind of distributional effort makes sense for most physical goods, because so many of them are both scarce and indivisible. The scarcity and indivisibility of physical goods means that distribution is inevitably a zero-sum game. And since human needs are packaged in discrete individuals, the needs and rights of individuals must be at the heart of any viable theory.

Introducing the Intellectual Commons

The problem with such an approach is that this is only half of the story.

We will consider three types of arguments for the moral significance of the intellectual commons. All three types appeal to intuitions about the nature of works and the nature of the human effort that creates them.

The first type of argument presents commons economics as a kind of "third way" between individual property rights and state property. The more formal members of this family are based on "new institutional economics," particularly the concept of an anticommons.[10] Other versions focus on the essential role an intellectual commons plays in making the creation of works possible.[11] The free and open source software movement is sometimes presented as an existence proof for the viability of commons-centered intellectual property.[12]

The second type of argument is favored by many political activists concerned with distributive justice. The hallmark of these arguments is the claim that some types of knowledge either cannot or should not be property. While the third-way arguments apply to all forms of IP, this second type applies most naturally to "useful knowledge": the domain of patents. And while third-way arguments tend to be formal, these community-based arguments tend to be driven by a broader moral framework that is shaped by specific situations. There is another difference as well: community-based arguments leave open the possibility of attacking the entire framework of intellectual property rights in a way that third-way arguments do not.[13]

Finally, there is the possibility of making arguments about intellectual property that are based in some kind of "deep ecology" that grants ideas and works some kind of independent and intrinsic moral significance. Whether it is is possible to construct any kind of rigorous argument of the basis of informational Platonism will be briefly discussed in the conclusion.

An example: The Free Software Movement

The prominent free-software developer Linus Torvalds is fond of the following argument: People write software for three reasons: to make money (or to gain status), to solve some problem, or "for the fun of it." In the first case, some kind of external compensation is necessary for the activity to be completed. But in the latter two cases, no external compensation is necessary for the activity to be complete. Giving the software away does not unsolve the problem it was intended to solve or take away the fun I had writing it. Given the existence of some infrastructure for distribution (which itself could be a public good), there is virtually no cost associated with sharing my work.[14]

The most modest way of reading this is as an argument for the rationality of sharing in some contexts. A community based on sharing does not necessarily require the renunciation of self-interest. The fact that I am writing this work using a system made up entirely of free software is evidence that such a community can flourish and produce great things. A recent study estimated that the Linux operating system required about 8000 man-years to construct. At prevailing market rates this represents an investment worth approximately 480 million dollars. Such a spectacular result is only possible because the creation of software requires so little in the way of physical resources. Linux is made from the "intense, devouring labor" Goldstein praises, but it is not made of "thin air." It requires the existence of what we will call a community of resources, which is a repository of skills, tools, and wisdom. It also requires the existence of what we will call a community of goals, a community whose norms of problem-solving and leadership support and reward individual effort. Finally, it also requires the existence of what we will call a community of trust, a community whose norms of sharing assure that self-sacrifice will not simply degenerate into the exploitation of the many by the few.

Conclusion

In this chapter we have had a great deal to say about the "property" part of IP. The next few chapters will take us through the history and institutional basis of intellectual property systems. It will be of necessity a rather long and circuitous journey, so I would like to begin by laying out the concepts we need to keep in mind:

- **Property Rights:** property can encompass both "exclusive" and "in-clusive" rights.[15] We must beware of the tendency to assume that property can only be understood in terms of exclusive rights, or in terms of the right to alienate. Commoners have neither exclusive access or the right to alienate, yet it seems reasonable to say that they have rights that a non-commoner does not have. We must also be mindful of the social dimensions of property ownership: even full liberal ownership is limited by the fact that we must share the world (and the consequences of our actions) with others.
- **Commons:** A commons can simply be the negation of the set of owned things. A commons understood in this purely negative sense functions as a precondition for original appropriation and nothing else. This is the commons that Locke saw at the beginning: "in the first Ages of the World, when Men were more in danger of being lost, by wandering from their Company, in the vast Wilderness of the Earth, than to be straitened for want of room to plant in."[16] The negative commons is both theoretically empty and inevitably transitory. It certainly cannot be relevant to any contemporary analysis of property.

  A positive conception of the commons is only possible where there are modes of ownership that are qualitatively different from full liberal ownership. The commons must be something that is both immune from dismemberment and available for use by individuals. A viable commons requires that there be ways to justly resolve individual claims without simply dividing the commons into individual pieces. In the terminology of James Tully, there must be a way to give each commoner a right to her due without assuming that this is equivalent to a right to her own.[17]
- **Community:** The twentieth century may be remembered as the era in which we discovered the limits of the natural world. It is no longer possible to assume that nature is so vast that nothing we can do to it

will have any lasting impact. We have also lost the "moral luxury" afforded by a boundless world: we can no longer assume that our own appropriation and consumption will always leave as much and as good for others.

The end of a boundless world thus imposes "common" moral constraints on individual property. These constraints involve the impact of our actions on other individuals and on the continued existence and health of the common world itself. A theory of property couched entirely in the language of individual rights cannot even express these constraints, let alone satisfy them.

The intellectual commons faces its own versions of these problems, since everything in the commons is a product of human effort and creativity. The intellectual commons needs new ideas and works the way the environmental common needs sunlight: without a sustaining energy, the system would cease to function. Changes in IP law and social institutions can hamper that flow as surely as dams and water pollution can reduce the value of a river.

- **Value:** The elegance and coherence of classical economics may tempt us to try to explain the value of property rights solely in terms of market value. This temptation must be resisted. Nor can we be satisfied with an explanation of value couched solely in terms of utility, since a thing has many different utilities simultaneously. We must also be aware of the tendency of economic theories to "externalize" costs that are not easily accommodated in economic theory.

- **Labor:** Labor provides the tightest moral link between an author and a work. In chapter 4 we will be looking in detail at labor as a criterion for granting property rights. The most important point to keep in mind about labor right now is the fact that labor is much more concrete and much easier to define than "creativity," or "genius," or the concept that an author creates a work *ex nihilo*. The rights of authors may be grounded as much in the work they do as in the "creative spark" that motivates the work.

IP regimes are almost impossible to understand a priori. Their current form is the product of a long and convoluted development that is intimately connected with the development of communication technologies. Nonetheless, there is a recognizable logic implicit in legal responses to technological changes. In the next chapters, we will explore the history of IP systems and their current institutional forms.

Notes

1. L. Ray Patterson and Stanley Lindberg, *The Nature of Copyright: A Law of User Rights* (Athens, GA: University of Georgia Press, 1991). Brian Martin, "Against Intellec-

tual Property," *Philosophy and Social Action* 21, no. 3 (1995): 7–22.

2. *NII White Paper*, 84.

3. Paul Goldstein, *Copyright's Highway: From Gutenberg to the Celestial Jukebox* (New York: Hill and Wang, 1996).

4. Foucault, "What is an Author?" 110.

5. Litman, "The Public Domain," 1008.

6. Price, *The English Patents of Monopoly*, 7.

7. The screenplay for the film *Toy Story* credits no fewer than seven writers.

8. Letter from Thomas Jefferson to Isaac McPherson, Aug. 13, 1813. The metaphor of the taper is a classic example from the poet Ennius which was also used by Cicero, Seneca, and Grotius.

9. Wendy Gordon, "A Property Right in Self-expression," 1539.

10. Elinor Ostrom, *Governing the Commons: The Evolution of Institutions for Collective Action* (New York: Cambridge University Press, 1990). See also Carol Rose, "Expanding the Choices for the Global Commons: Comparing Newfangled Tradable Allowance Schemes to Old-fashioned Common Property Regimes," *Duke Environmental Law and Policy Forum* 10 (2000): 45–72.

11. Litman, "The Public Domain," 1000.

12. Chris Dibona, ed. *Open Sources: Voices From the Open Source Revolution* (San Francisco: O'Reilly Associates, 1999).

13. Vandana Shiva, *Biopiracy: The Theft of Nature and Knowledge* (London: South End Press, 1997).

14. Linus Torvalds and David Diamond, *Just For Fun: The Story of an Accidental Revolutionary* (New York: HarperBusiness, 2001), 226-227.

15. Tully, *A Discourse on Property*, 61.

16. Locke, *Two Treatises of Government*, II 36.

17. Tully, *A Discourse on Property*, 61.

3

An Overview Of Intellectual Property

> Another, more general manifestation of the same trend has been the growing
> power of the phrase "intellectual property." Before the Second World War, use
> of the phrase as a shorthand for copyrights, patents, trademarks and trade se-
> crets was rare. Since that time it has become steadily more common. Today it is
> the standard way for lawyers and law teachers to refer to the field. Why does
> the popularity of the term matter? The answer—as the Legal Realists recog-
> nized long ago—is that legal discourse has power. Specifically, the use of the
> term "property" to describe copyrights, patents, trademarks, etc. conveys the
> impression that they are fundamentally "like" interests in lands or tangible per-
> sonal property—and should be protected by the same generous panoply of
> remedies. (Fisher, "Growth," 9)

We will make a quick tour through the four main forms of intellectual property:
trade secrets, patents, trademarks, and copyright. The reader should note that the
concepts have considerable room for overlap: a computer program could begin
as a trade secret and later become patented or copyrighted. Intellectual property
forms differ somewhat in subject matter, but differ profoundly in terms of duration, procedural rules, and length of term.

The object of IP ownership is not a physical object, but a bundle of rights
connected with the economic value of a work. The practical form of these rights
has been to focus on copying (in the case of copyright and trademark) and using
(in the case of patents).[1]

There are two ways of making the distinction clear. The first is to focus on
the author's property rights in an unpublished manuscript and his property rights
in a published work. The author has full liberal ownership of his unpublished
manuscripts: this ownership is the same as his ownership of his shoes or his furniture. Once the manuscript has been published or registered for copyright, the
author's rights are determined by a completely different legal regime. The second way of illustrating this point is to consider the difference between a concrete
and abstract conception of works. It is sometimes argued that the development of
copyright law and the emergence of author's rights was a transition from a concrete to an abstract view of what a "book" is. If books are material objects that
are produced at great labor and expense by publishers, it makes sense that "copyright" should be a right possessed by publishers. If books are the type created by
great labor and expense by the author, and the "copies" made by publishers are
mere tokens of that type, it makes sense that "copyright" should be a right pos-

sessed by authors (or at least that both should have rights with regard to the work).

Copyrights

For many copyright represents the paradigmatic form of intellectual property. Copyright is intended to guarantee the author of a work certain rights, both economic and "moral." The economic rights are the benefits gained from exclusive control: the work cannot be published without the author's consent. Moral rights are non-economic rights derived from the nature of authorship: "The author's moral rights principally protect his entitlement to object to any unauthorized modification of his work, and to claim authorship of it. The work must not be modified without the author's consent, at any rate in a way prejudicial to his honour or reputation."[2]   U.S. copyright law recognizes moral rights of authorship only for visual works.[3]

Copyright is the most technology-driven area of intellectual property law. The original subject-matter of copyright was literary works, but over the last three hundred years the scope of copyright law as grown to include maps, sheet music, all forms of recorded performances, photographs, player piano rolls, computer programs, and the layout masks for integrated circuits.

The term of copyright protection has increased as well: in 1710, the Statute of Anne granted an initial copyright for fourteen years that could be renewed once. The current term of copyright is the author's lifetime plus seventy years.[4]

At the same time, the formal requirements for copyright have been greatly reduced: even registration with the Library of Congress is no longer required (though registration grants additional rights important in litigation). In principle, any written work, no matter how trivial, is automatically covered by copyright (love letters, grocery lists, to-do lists, etc).[5]

Copyright law explicitly makes a distinction between the content of a work (its "ideas") and its concrete form of realization ("expression"). Copyright is intended to protect expressions, not ideas. This distinction has both ethical and pragmatic motivations.

The idea–expression distinction has been strained by the development of performances and recordings. The same musical work can be expressed as sheet music, a player piano roll, a digital or analog recording, or a computer program. Even beyond these concerns, almost every case of copyright infringement that doesn't involve bald plagiarism requires drawing the idea–expression boundary for a particular work. Mark Rose describes being called as an expert witness in a case of copyright infringement, where a science fiction writer was accusing the screenwriter of a television show of copyright infringement.[6]

The writer had published a story in which a policeman was forced to accept a robot as his partner even though he feared the robot would take his job. The TV show was also based on the same premise, but differed in various details (e.g., the robot in the story looked like a vacuum cleaner, while the robot in the TV series was a virtually perfect duplicate of a human being).

The idea/expression distinction is particularly important for copyright because copyright does not cover the independent creation of an identical work. If the writers of the TV show described above can give some evidence of independently coming up with the "policeman-forced-to-work-with-robot-he-distrusts" idea, then their work isn't an infringement of the science fiction writer's copyright.

The most striking example of the independent-creation defense is the process of clean-room reverse engineering of software.[7]

In the clean-room process, two teams of engineers work to copy an existing program. The first team "reverse engineers" the product by disassembly and exhaustive testing. They then produce a written specification that totally describes the behavior of the program. The second team takes the specification and writes a completely new program that exactly embodies the specification. The second team is "clean" because they never see the original product, only the specification. The end result is a program that is a functional duplicate of the original program, but that is not a "literal" copy. The first team takes the legally protected expression of the original program and converts it into a legally unprotected " idea" (the specification). The second team then takes the idea and converts it back into a new legally protected expression.

Reverse engineering seems at first blush like theft. However, it should be noted that the process requires as much work as the development of the original program (if not more). Pirates usually want a free ride on the efforts of others, not a long and expensive development effort. Not all reverse engineering efforts are designed to exactly duplicate a product: most are intended to allow the development of programs that can "plug into" or operate compatibly with an existing program. Some reverse engineering isn't commercial at all: the BSD Unix distribution was developed by reverse engineering after ATT imposed large licensing fees on a previously free operating system.[8]

Copyright is also limited by the doctrine of "fair use." Certain uses of copyrighted material do not require the permission of the copyright holder. These uses include "criticism [this includes parodies], comment, news reporting, teaching (including multiple copies for classroom use), scholarship, or research."[9] However, not all such uses are considered "fair." Four other criteria are relevant:

- The purpose and character of the use (commercial or not)
- The nature of the work (copying a textbook to avoid buying one is not considered fair use in a class, since students would normally be expected to purchase textbooks in such a setting)
- The amount of the work copied (the more copied, the less fair)
- The effect of the use upon the potential market for, or value of, the copyright (a group of libraries can't "share" a journal subscription through copying).

Exceptions for fair use and parody are motivated by the ultimate rationale for copyright: "the advancement of science and the useful arts." All of the uses specially identified as fair involve placing a work into the context of existing knowledge, rather than appropriating it for economic gain.

There are other reasons for asserting the idea/expression boundary and the requirement of originality. Allowing the ownership of ideas would ultimately impede the advancement of knowledge, especially for the long terms allowed by copyright law. The requirement of originality is at least partially intended to prevent someone from gaining intellectual property rights by simply being the first one to ask for them. I shouldn't be able to copyright a work simply because its author never chose to.

Patents

Patents are government-granted monopolies on the use of certain kinds of inventions. The term "invention " must be understood in a broad sense: not just a machine, but any of the following:

- a process (e.g., a way of making pre-faded jeans)
- a machine
- a manufacture (relatively simple objects that don't have working or moving parts e.g., a pencil)
- a composition of matter (e.g., vulcanized rubber, genetically modified organisms)
- an improvement of an existing idea (though this only covers the improvement, not the original idea)

In order to qualify for patent protection, an invention has to be useful, novel, and non-obvious. The condition of non-obviousness was added to the law in 1952: in many ways it serves a function similar to that of originality in copyright law. Fred Warshofsky explains it as follows: "In other words, simply the fact that an invention has not actually been invented before does not mean it was not conceivable by someone skilled in that particular field who could have invented it relatively easily had he or she tried. Such an invention would be novel, thus meeting one of the conditions of patentability, but also obvious to such a person and therefore not patentable."[10]

Patents are intended to reward an inventor for a significant effort to add something new, not simply for being the first to claim something previously unclaimed. In addition, it has to be an invention, not a discovery: laws of nature, mathematical theorems, and naturally occurring substances are (supposedly) unpatentable.

The patent application process involves submitting an application that gives complete, detailed instructions of the invention, a review of prior inventions intended to prove that the invention is truly novel, and a set of "claims" exactly describing what aspects of the invention are to be protected by the patent. A patent examiner reviews the application and can deny it or send it back for revision. If two inventors simultaneously file an application for the same invention, a special administrative review called an "interference" is conducted. The entire process is long, arduous, and can be extremely expensive. Even if a patent is granted, it can be legally challenged and possibly invalidated. These legal battles can reach apocalyptic proportions: a patent dispute between Kodak and Polaroid about instant film took from 1976 to 1986 to resolve and ended up costing Kodak almost a billion dollars in damages.[11]

Patents are the strongest form of intellectual property protection. Unlike copyright, which protects expressions, patents can protect ideas themselves. Patented inventions cannot legally be reverse engineered. Anyone who uses the RSA encryption algorithm without permission is infringing RSA's patent, regardless of what specific form their program takes. Even the *totally independent creation* of the algorithm by someone else would not give them the right to use it without permission from RSA.

Because of their power, patents also have the shortester term in force: a maximum of twenty years from the date of first filing, compared with up to ninety-five years for copyrights and no time limit for trademarks and trade secrets.

Trade Secrets and Trademarks

In addition to copyrights and patents, IP law recognizes two other forms of ownership: trade secrets and trademarks. Though we will have little to say about them in what follows, it is appropriate to give a basic description.

Trade secrets are the original form of intellectual property. According to the Uniform Trade Secrets Act,

> "Trade secret" means information, including a formula, pattern, compilation, program, device, method, technique, or process that: (1) derives independent economic value, actual or potential, from not being generally known to, and not being readily ascertainable by proper means by, other persons who can obtain economic value from its disclosure or use, and (2) is the subject of efforts that are reasonable under the circumstances to maintain its secrecy.[12]

Trade secret law protects information whose values is *derived from its secrecy*. That secrecy must be protected by reasonable efforts at security, not simply by default.

Trade secret law is profoundly different from the other branches of IP law. It is the only form of IP protection that doesn't require prior disclosure of the works to be protected (which would be self-defeating). Trade secrets are also the only form of IP without some central certifying authority. Trade secrets are not based on originality or utility, but simply on their economic value as secrets. Disputes are fought out against the background of contract law (e.g., nondisclosure agreements) and unfair competition law (e.g., industrial espionage).

Trade secrets have traditionally been considered the weakest form of intellectual property protection, since they can only be asserted when they have already been breached. Ironically, attempting to enforce protection of trade secrets can undermine their nature as secrets. Conceptually, they are most interesting in that they seem to bear some analogy to the concept of personal privacy rights.[13]

Trademarks

Trademarks are names or symbols associated with a particular product or service. As defined by the Patent and Trademark Office, "A Trademark is either a word, phrase, symbol or design . . . which identifies and distinguishes the source of the goods or services of one party from those of others."[14]

Unlike patents and copyrights, the stated rationale of trademark is purely commercial. Trademark facilitates the creation of product brands and prevent competitors from misleading consumers with falsified packaging.

U.S. trademarks are administered by the U.S. Patent and Trademark Office, but trademark rights can exist independently of an approved registration. Use of the mark in commerce is enough to establish *de facto* trademark rights, though registration provides additional rights. U.S. trademark law is unique in *requiring* the use of a mark before it can be registered; most other countries allow registration before use.[15]

Applications for trademark registration are evaluated by comparing them to existing trademarks and determining if they would introduce a substantial risk of confusion with the existing marks. Trademark law distinguishes between "Strong" and "Weak" marks: common words and names are considered weak, while newly coined terms (e.g., "Zoloft" and "Trigon") are considered strong marks. The distinction between strong and weak marks plays a role similar to that of novelty in patent law.

Trademarks have a term of ten years, but can be renewed indefinitely. Failure to renew usually means loss of trademark rights: the rule in trademark law is "use it or lose it."

The world of intellectual property is vast and complicated: this survey has only scratched the surface of a complex and subtle discipline. In the next chapter, we will begin to analyze the moral justification for IP in detail. We will begin by looking at two important moments in the history of copyright. At these crucial junctures we can see both careful analysis and the blinding effect of preconceptions.

Notes

1. These practical forms are not inevitable: in "The Exclusive Right to Read," Jessica Litman argues that an exclusive focus on copying is inappropriate to digital media and that the true focus of copyright law should be the author's right to benefit economically from publishing his work. Presumably this could be accomplished by focusing on licensing rather than the mere act of copying.

2. "Proposal for a European Parliament and Council Directive on the Harmonization of Certain Aspects of Copyright and Related Rights in the Information Society," Technical Report, European Union, 1997. Chapter II, section VII.

3. Even this concession was motivated only by the need to harmonize U.S. copyright law with the Berne convention.

4. "Duration of Copyright," circular 15a, Library of Congress Copyright Office, April 1999.

5. Litman, "The Public Domain," 974.

6. Rose, *Authors and Owners,* 134.

7. Fred Warshofsky, *Patent Wars: The Battle to Own the World's Technology* (Hoboken, NJ: John Wiley and Sons, 1994.) 123.

8. Peter Wayner, *Free for All: How Linux and the Free Software Movement Undercut the High-Tech Titans* (New York: HarperCollins, 2000), 37-52.

9. Library of Congress, "The Copyright Law of the United States." Circular 92, Library of Congress Copyright Office, 1998.

10. Warshofsky, *Patent Wars,* 51.

11. Warshofsky, *Patent Wars,* 69-88.

12. Stephen Elias, *Patent, Copyright & Trademark* (Berkeley, CA: Nolo Press, 1999), 64.

13. Margaret Jane Radin has argued that the moral foundation of IP rights is an individual's right to keep secrets. See "Property and Personhood," *Stanford Law Review* 34 (1982): 957. See also Lynn Sharp Paine, "Trade Secrets and the Justification of Intellectual Property," *Philosophy and Public Policy* 20 (1991): 247.

14. "Basic Facts About Registering a Trademark," Technical Report, U.S. Patent and Trademark Office, 1997.

15. Elias, *Patent, Copyright & Trademark,* 338.

4

Two Moments in the History of Copyright

The *Donaldson v. Becket* decision of 1774 was a crucial moment in the history of copyright. By the narrowest of margins, the English House of Lords rejected an attempt to absorb copyright into common-law property. The decision kept open the idea that IP rights are inherently limited by their dual purpose. Reviewing the debates over the decision gives us a window into how both proponents and skeptics reasoned about a natural right of authors.

The U.S. Supreme Court decision on the *Eldred v. Ashcroft* case in January of 2003 may or may not have the same historical importance: only time will tell. Yet there are fascinating parallels between the cases. In both, an attempt to impose consideration of natural rights was rejected by the courts. In *Donaldson*, a natural right of authors was rejected; in *Eldred*, much more indirectly, a natural right of users was rejected. The earlier debate is colored by the invocation of natural rights on both sides: the latter, by a studious attempt on both sides to argue only in terms of legal precedent and rational choice theory. But though the *text* of the majority decision in *Eldred* generally embraces this severe, functional style, the *footnotes* tell a quite different story. Here, Justice Ginsburg, writing for the majority, engages in a viciously polemical running battle with the two dissenters (Justices Breyer and Stevens). The dissenting opinions, especially Breyer's, reflect the same sense of frustration (even bitterness) that is so present in the majority opinion. One has the sense that Eldred had the misfortune to stumble into a deep-seated and very bitter family argument. The debate reveals a profound difference of opinion on the current direction of copyright law: and that difference is motivated by an equally profound disagreement over the purpose of copyright law.

When the Statute of Anne ended the Stationer's monopoly on printing in 1710, they went to court to argue that their copyrights were "common-law" property that had no term of ownership. The courts finally ruled against them in 1774, bringing copyrights fully under government control.[1]

The Stationer's gradual decline throughout the eighteenth century was matched by gradual elevation of the legal status of authors. [2] The Statute of Anne made authors rather than publishers the original owners of copyrights, giving them a central economic role in publishing for the first time. By the end of the eighteenth century, the *ideology* of copyright was dominated, and is still dominated, by the centrality of the author.

The Statute of Anne and "the Battle of the Books"

Copyright was originally a right of publishers, not one of authors. The transition to an author-based theory began in 1710, with the passage of "An Act for the Encouragement of Learning, by vesting the Copies of printed Books in the Authors or Purchasers of such Copies, during the Times therein mentioned," generally known as the Statute of Anne (see appendix C for the complete text).

The Statute of Anne laid the foundation for all copyright law to follow. On the surface, the Statute seems to give the Stationers what they wanted: their role in registering copyright is continued, and their existing copyrights are grandfathered in. However, the role left is ultimately a hollow one: they can be fined for refusing to register a work, and they can be bypassed completely if necessary (section III). Disputes about copyrights are to be resolved in the regular courts instead of in the Stationer's Company.

A more profound change comes in section II. The Statute allows anyone to purchase a copyright from an owner, not just the Stationers. The Statute also puts no restrictions on who can print books: this silence effectively broke the Stationer's monopoly. To add insult to injury, section IV sets up a system of price controls for books (though there is no evidence it was ever enforced).

The most radical change of all is the original vesting of copyright in authors. Authors are vested with a copyright which they can assign to others, and that lasts for fourteen years (the same term granted to patents in the reform law of 1624). Living authors can renew a copyright for a second fourteen year term. This gave authors a "residuary interest" in their works: they can renegotiate assignment or withdraw it at the end of the first term. All works became immune from copyright at the end of twenty-eight years, or at the death of the author. Though authors received more property rights than ever before, they were still significantly less than full liberal ownership.

L. Ray Patterson has argued that the Statute of Anne's focus on authors was merely a tactic to attack the Stationer's monopoly, and should not be read as motivated by some sense of moral obligation:

> The monopolies at which the statute was aimed were too long established to be attacked without some basis for change. The most logical and natural basis for the changes was the author. Although the author had never held copyright, his

interest was always promoted by the Stationers as a means to their end. Their arguments had been, essentially, that without order in the trade provided by copyright, publishers would not publish books, and therefore would not pay authors for their manuscripts. The draftsmen of the Statute of Anne put these arguments to use, and the author was used primarily as a weapon against monopoly.[3]

Adrian Johns and Mark Rose have both questioned Patterson's derogation of author's rights. Johns argues that there is no reason to attribute the degree of cynicism that Patterson's claims seem to require: "In fact, the Stationers had developed their own concept of civil order, of which 'copy' was a manifestation . . . the Stationer's civility has often been mistaken, then and now, for bald rapacity."[4] Rose and Johns see the Statute of Anne as a reflection of the emerging professionalization of authorship and the development of a literary theory that valued creation over reinterpretation.

Making the author the original source of copyright proved to be a very effective weapon indeed. It also has at least one surprising consequence: the creation of a "public domain." If ownership of IP originates solely in authors, then existing works whose authors are long dead or unknown are not owned by anyone. IP ownership is also further circumscribed by imposing a limited term on ownership.

The Statute was a massive blow to the Stationer's Company but not a fatal one: it grandfathered in their existing copyrights for a period of twenty-one years, and also continued existing printing patents. When the grandfather copyrights began to expire, the Stationers began to sue "infringers" in civil court, in an effort to establish some form of common-law (and permanent) copyright. These legal battles continued until 1774, when an appeals court explicitly stated that the only copyright is a statutory one.

The Stationers were to change direction after this crushing defeat. With the decline of censorship, their monopoly on printing lost any larger legal justification. The stationers were forced to lobby in defense of their existing copyrights, while bemoaning the devastating effects of literary piracy on their widows and orphans. An attempt to pass a copyright law (as opposed to a censorship law) failed in 1707, but finally succeeded in 1710, ultimately producing the Statute of Anne. (See appendix C for the text of the statute.)

The decline of censorship left the Stationers with a need to find a new justification for their monopoly power. The Stationers realized that they could only defend their own economic interests by supporting the property rights of authors. This support had always been a part of their rhetoric, and drew on the longstanding metaphor of the author as the "master" of her work.

Strictly analogical arguments are explicit in many early claims of author's rights.

It is humbly conceived First, That the Author of every Manuscript or Copy hath (in all reason) as good right thereunto, as any Man hath to the Estate wherein

> he has the most absolute property; and consequently the taking from him the
> one (without his consent) will be equivalent to the bereaving him of the other,
> contrary to his Will. (The Stationers v. The Patentees [1666], argument of ap-
> pellant.[5])

If we accept the analogy, perhaps we can argue from the exclusive rights of real
property to the control rights of IP. Though couched in the language of sixteenth-century absolutism, the following quote makes the position clear:

> In the same way, the author of a book is wholly its master, and as such he can
> freely do with it what he wills; even keep it permanently under his control as he
> might a slave; or emancipate it by granting it common freedom; giving that
> freedom either purely and simply, without holding back anything, or else im-
> posing some limits, by a kind of patronage, so that no one but he will have the
> right to print it except after a certain time.[6]

The Stationers could agitate for the property rights of authors because they still
expected authors to sell these rights to their publishers. The author's full and perpetual ownership of work would then legitimize the publisher's claim of full and
perpetual ownership. As long as the Stationers controlled the economic and administrative systems of copyright, their monopoly would be secure.

The Stationers had new allies in 1709—professional authors like Daniel Defoe and Joseph Addison. Addison presented the following elevated defense in
December 1709:

> All Mechanick Artizans are allowed to reap the Fruit of their Invention and In-
> genuity without Invasion; but he that has separated himself from the rest of
> Mankind, and studied the Wonders of the Creation, the Government of his Pas-
> sions, and the Revolutions of the World, and has the Ambition to communicate
> the Effect of half his Life spent in such noble Enquiries, has no property in
> what he is willing to produce, but is exposed to Robbery and Want, with this
> melancholy and just Reflection, That he is the only Man who is not protected
> by his Country, at the same Time that he bests deserves it.[7]

Defoe was a bit more blunt:

> A Book is the Author's Property, 'tis the Child of his Inventions, the Brat of
> his Brain; if he sells his Property, it then becomes the Right of the Purchaser; if
> not, 'tis as much his own, as his Wife and Children are his own–But behold in
> this Christian Nation, these Children of our Heads are seiz'd, captivated, spirit-
> ed away, and carry'd into Captivity, and there is none to redeem them.[8]

Donaldson v. Becket

Booksellers and authors began a series of lawsuits in 1735 (the year that grandfathered Stationer's copyrights expired) attempting to establish the existence of a common-law perpetual copyright. They achieved their goal with *Millar v. Taylor* in 1769, when the Court of King's Bench ruled 3-1 that there was a perpetual common-law copyright.[9]

The issue was revisited in 1774, when the House of Lords heard an appeal in the case *Donaldson v. Becket.* "Immediately at issue was whether Alexander Donaldson, a Scottish bookseller who had built a successful business on cheap reprints of the classics, had acted as a pirate when, six years before, he had published an edition of James Thomson's *The Seasons*, a work for which Thomas Becket and a group of other London booksellers and printers claimed the copyright."[10] The appeal was heard in the House of Lords before the twelve most senior judges in England (from the Courts of King's Bench, Common Pleas, and Exchequer). After the judges ruled, the Lords themselves voted. The judges ruled for Donaldson 6-5 (Judge Mansfield having recused himself because his involvement in *Millar v. Taylor*), and rejected perpetual common-law copyright, with the Lords following suit.

The official transcript of the Lords' debate on literary property deserves some close attention because it tells us so much about how legal copyright was understood by those within two generations of its founding.[11] As we shall see, these arguments also mirror the arguments we will be studying in *Eldred v. Ashcroft.*

Attorney General Thurlow took a cynical view of the whole process, particularly the focus on author's rights. "The booksellers . . . had not, till lately, ever concerned themselves about authors, but had generally confined the substance of their prayers to the legislature, to the security of their own property; nor would they probably have, of late years, introduced the authors as parties in their claims to the common law right of exclusively multiplying copies, had not they found it necessary to give a colourable face to their monopoly." Chief Justice De Gray also argued a similar point in his opinion:

> If such a right existed at common law, and it remained unimpeached by that statute, why that anxiety in authors and booksellers afterwards to obtain another sanction for their property? whence those different applications to parliament, in the years 1735, 1738, 1739, for a longer term of years, or for life, in this kind of property, and afterwards to get an act to prohibit the liberty of printing books in foreign kingdoms, and sending them back again. The truth is, the idea of a common-law right in perpetuity was not taken up till after that failure in procuring a new statute for an enlargement of the term. If (say the parties concerned) the legislature will not do it for us, we will do it without their assistance; and then we begin to hear of this new doctrine, the common law right . . .

Lord Camden argued that the Precedents cited by the respondents are the *exact opposite* of what one would expect for a common-law argument: "The arguments attempted to be maintained on the side of the Respondents, were founded on patents, privileges, Star chamber decrees, and the bye laws of the Stationers' Company; all of them the effects of the grossest tyranny and usurpation; the very last places in which I should have dreamt of finding the least trace of the common law of this kingdom." The respondents' precedents instead highlight the real basis of IP rights: "The manner in which the copy-right was held was a kind of copyhold tenure, in which the owner has a title by custom only, at the will and pleasure of the lord. The two sole titles by which a man secured his right was the royal patent and the licence of the Stationers' Company; I challenge any man alive to shew me any other right or title."

Several of the justices wondered why this "common-law" right of authors was just being argued for. If it really was a common-law right, why hadn't anyone noticed it before? Why didn't authors attack the Statute of Anne for taking away their natural rights?

> But, though the press was ever an object both to legal and usurping princes, yet in no regulations respecting it was a common law right in books noticed in the most distant manner; yet had such right existed, we surely must have heard of it, particularly as some of the British princes were authors . . . The statute vested a property in the author for 14 years, "and no longer." Sir John [Dalrymple] asked, why the phrase "and no longer" was adopted? Admitting the respondents right in their notions about a common law property, a claim so founded must vest the property in the owner for a perpetuity: How then could this statute be called, as it is, "An Act for the better encouragement of learning". . . Upon the supposition of a common law right, the statute curtailed, instead of enlarging an author's privileges; it vested nothing in him but what he had before; it secured nothing to him but what he was previously secured in by the common law; and in the place enjoying a property transmissible in perpetuity to his heirs, he enjoyed one for 14 years only.

The justices who rejected common-law rights made a great deal of the parallel between authors and inventors. Their purpose was to undercut the claim that an author's intellectual labors vested him with a natural property right. Their argument was simply that inventors claimed no *natural* property rights in their inventions even though their inventions required as much intelligence and commitment as authors showed.

> The maker of a time-piece, or an orrery, stood in the same, if not in a worse predicament, than an author. The bare invention of their machines, might cost them twenty of the most laborious years in their whole life; and the expence to the first inventors in procuring, preparing, and portioning the metals, and other component parts of their machines, was too infinite to bear even for a moment the supposition that the sale of the first orrery recompensed it. And yet no man would deny that after an orrery was sold, every mechanist had a right to make another after its model. The baron considered a book precisely upon the same footing with any other mechanical invention. In the case of mechanic inventions, ideas were in a manner embodied, so as to render them tangible and visi-

ble; a book was no more than a transcript of ideas; and, whether ideas were rendered cognizable to any of the senses, by the means of this or that art, of this or that contrivance, was altogether immaterial: yet every mechanical invention was common whilst a book was contended [by the stationers] to be the object of exclusive property!

This argument might strike a modern reader as puzzling, given the hundreds of thousands of patent applications processed by the USPTO every year. This sense of bewilderment is a sign of how much the landscape of intellectual property has changed since 1774. At the time of this decision, the British government was only granting about thirty patents a year (up from an average of five a year in the first half of the eighteenth century).[12]

The vast majority of inventors never even considered seeking a patent for an invention. The first patent law in the United States specified that the Secretary of State would personally evaluate all patent applications. Many of the justices saw granting a patent as an extraordinary privilege granted by the state in the interest of promoting industrial progress. There was no "right" to patent, anymore than today someone has a "right" to win the Nobel Prize.

Chief Justice De Gray's final argument against the idea of a common-law copyright points to the many forms of creative activity for which there was no claim of common-law right.

Abridgments of books, translations, notes, as effectually deprive the original author of the fruit of his labours, as direct particular copies, yet they are allowable. The composers of music, the engravers of copper-plates, the inventors of machines, are all excluded from the privilege now contended for; but why, if an equitable and moral right is to be the sole foundation of it? Their genius, their study, their labour, their originality, is as great as an author's, their inventions are as much prejudiced by copyists, and their claim, in my opinion, stands exactly on the same footing; a nice and subtle investigation may, perhaps, find out some little logical or mechanical differences, but no solid distinction in the rule of property that applies to them can be found.

His point is that there is no obvious "natural" difference between writing a book and the other unprotected activities. Ironically, his statement can now be seen as a vision of the future of intellectual property law.

Eldred v. Ashcroft and the Battle for Copyright Term Extension

The 1790 copyright act set the term for U.S. Copyright at fourteen years, renewable once (the same term specified by the Statute of Anne). A lot happened to American Copyright law in the two hundred years that followed:

The 212-year history of copyright in the United States is a history of constant expansion of rights. The first copyright act of 1790 protected only maps, charts and books, but subsequent acts have added dozens of new creative works: prints in 1802, music in 1831, photographs in 1865, dramatic works in 1870, paintings, drawings and sculpture in 1870, movies in 1912, sound recordings in 1971, dance in 1976, computer programs in 1980, architectural works in 1990, and boat hull designs in 1998. During the same period, Congress has expanded the scope of rights to include not only the rights to copy and distribute works, but also the exclusive rights to perform certain works publicly in 1856, to create certain derivative works in 1870, to display certain works publicly in 1976, to preserve the integrity of and to claim authorship of certain visual works in 1990, to create and distribute recordings of live musical works in 1994, and to protect works by technological protection measures and the encoding of copyright management information in 1998. Congress has also expanded copyright by extending the term of copyrights in 1831, 1909, 1976, and 1998; and by eliminating the formalities of notice, registration, deposit and renewal, primarily in 1988.[13]

Congress increased the total possible term to forty-two years in 1831 and fifty six years in 1909. The 1976 Copyright act abandoned fixed terms, setting copyright term to the life of the author plus fifty years. The most recent extension was the Sonny Bono Copyright Term Extension Act of 1998 (CTEA), which expanded the term to the life of the author plus seventy years. The primary justification given for the extension was to coordinate American and European copyright terms, though the extent of this harmonization has been questioned. Two features particularly galled the law's critics: that it gave a benefit that would only be realized long after the death of the author, and that it was also extended retroactively to current copyright holders. While the heirs of a currently living writer might receive a boon in the distant future, Disney Corporation and Universal Studios get another twenty years, starting today, to profit from their copyrights of the 1920s.

Arguments against the CTEA took many forms, and we will quickly highlight a few of them:

1. *The CTEA is defective as an instrument of international harmonization*: The CTEA makes some copyright terms longer than their equivalents under the Berne Convention, and also leaves some shorter. It continues to use some fixed-year terms, which are irreconcilable with the "life plus" rules of the Berne Convention. The claim for harmonization is further undercut by title II of the bill, the "Fairness in Music Licensing Act" which was added to the bill in committee. This law has been *already* been ruled in violation of the TRIPS agreement by the World Trade Organization, which is considering penalties.[14]

2. *The CTEA does not create a significant incentive for the production of new works*: The twenty years added to copyright terms will only be

realized after currently living authors have been dead for fifty years. This "enhancement" of the incentive has virtually no present value. However, it is very valuable indeed to the corporations and heirs holding the copyrights of works from the 1920s that would otherwise enter the public domain. Critics point out that these owners have already had the benefit of decades of royalties. Nor is there any evidence whatever that the current owners are as creative as the original authors, or that the current owners would insist on investing this new income into the creation of new works.

3. *The CTEA represents an attempt to make copyright terms effectively unlimited*: critics claimed to see a pattern in a twenty-year retroactive extension of copyright being enacted around twenty years after the Copyright Act of 1976. Some argued that Congress was attempting to render "limited times" meaningless by a potentially unlimited series of extensions to copyright term. Their anxieties were further heightened by statements made during the hearings in support of perpetual copyright.[15]

4. *Explicit Social Contract Arguments*: critics made two broad arguments against the CTEA on social-contract grounds:

   • It is unfair to increase copyright protection on existing works because this imposes a new public cost without any new public benefit. This would be tantamount to Congress extending a lease on public lands for a century without changing the current rent.[16] Retrospective increases in copyright term are never justified because they never play any significant role as incentives for the creation of new works.

   • Since the purpose of copyright is to promote the progress of science and industry, Congress needs to make a deliberate and thorough attempt to demonstrate that a new grant of rights to authors is counterbalanced by a new benefit to the public.

In 1999 Eric Eldred, a publisher who specializes in public-domain works sued the government to block the implementation of the CTEA. Eldred's suit was dismissed by the circuit court and also rejected by the court of appeals. The case was finally decided against Eldred by the Supreme Court in a 7-2 decision on January 21, 2003.

The final decision (by Judge Ruth Bader Ginsberg) makes compelling reading, along with the vast number of *amicus* briefs filed by both sides.[17] Virtually all prominent copyright scholars wrote articles supporting or attacking copyright extension: in many ways the controversy over the CTEA was greater than that over the much more radical Digital Millennium Copyright Act. The heightened interest was a sign that a new skepticism about copyright was finally beginning to leave the pages of law reviews and take on political shape.

Eldred v. Ashcroft resembles *Donaldson v. Becket* in being an attempt to use judicial review to overturn legislation. However, *Donaldson* was in a sense significantly less burdened by precedent than *Eldred*. *Donaldson* also had the advantage of an argument that was in harmony with the moral intuitions of much of its audience. No matter how good Eldred's arguments against retroactive term extensions were, they were inevitably dashed on the rock of precedent: every extension of copyright term in the U.S. has, in fact, been both prospective and retroactive. Given that Eldred's arguments against retroactivity risked overturning virtually all existing precedents, it is unsurprising that they were so soundly rejected.

The petitioners saw their lawsuit as an attempt to reinforce an interpretation of copyright as involving a *quid pro quo* between copyright owners and the public. "These extensions have overturned the constitutional balance between granting limited monopolies to authors and sustaining the vitality of the public domain. The issue in this case is whether this constitutional balance is to have any force in the context of changes to the copyright term."[18] The majority rejected this argument on several grounds:

- *Copyright is justified as a system, not as individual grants*: "The 'constitutional command' we have recognized, is that Congress, to the extent it enacts copyright laws at all, create a 'system' that 'promote[s] the Progress of Science.' *Graham, 383 U. S., at 6. 18.*" The grant of copyright is not an individual contract between an author and the government, but the admission of a work into a legally protected class. The system is justified *in toto*, and thus does not have to justify every individual change to copyright term.
- *The primary purpose of the copyright system is to create incentives for the creation of works*: Quoting *Mazer v. Stein*, the majority argued that copyright exists in order to create incentives for authors to share their works. Providing these incentives *automatically* promotes "the progress of science and the useful arts." Changing circumstances will sometimes require changing incentives, including longer terms.
- *The twenty-year extension provides a real incentive for the production of works*: The fact that it may be a very *slight* incentive is irrelevant as long as it is, in fact, a real incentive.
- *No explicit "quid pro quo" is required for Congress to give authors and owners more benefits*: Since copyright is justified systematically, and since giving authors more incentives automatically promotes dissemination of works, there is no reason to demand something back from the authors every time copyright is changed.
- *"Limited Times" means "not perpetual" and nothing else*: "Petitioners' argument essentially reads into the text of the Copyright Clause the command that a time prescription, once set, becomes forever

'fixed' or 'inalterable.' The word 'limited,' however, does not convey a meaning so constricted. . . . Thus understood, a time span appropriately 'limited' as applied to future copyrights does not automatically cease to be 'limited' when applied to existing copyrights." (Majority opinion, 8).

- *Pattern? What pattern?* The majority saw no attempt to achieve perpetual copyright by installments, since the law had a clear purpose (international harmonization) that transcended a mere handout to owners of old works.
- *Problem? What problem?* The majority emphatically rejected any suggestion that there should be any kind of routine judicial oversight of copyright legislation, on either First Amendment or Fourteenth Amendment grounds. The only things Congress *can't* do with copyright law is either (a) impose an explicitly perpetual term or (b) reimpose copyright on a work already in the public domain.

Common Themes

Those agitating against the CTEA saw Eldred's legal challenge as an opportunity to challenge the enormous expansion of copyright since the 1976 copyright reform. Many also saw the case as a chance to establish a new paradigm for IP law that weighs the social costs of awarding IP rights. Such dreams were dashed by the majority, who insisted on reading the issues at hand in the narrowest possible sense. Perhaps the narrowest interpretation of all was given to the specification of "limited times," which now means nothing more than "not perpetual." The logic of the majority decision is driven by the assumption that the only purpose of copyright law is to create incentives for authors, and that if incentives are created or enhanced the system is working. It is entirely up to Congress, not the courts, to determine appropriate levels of incentives. The CTEA's stated purpose of harmonizing U.S. and international copyright terms gave the court permission to sidestep any question of whether or not the extension was systemically justified. Whether the next extension can be dealt with so simply is another matter.

The plaintiffs were ultimately undone by their argument that retroactive term extensions are unconstitutional because there is no reasonable way to "incentivize" the creation of works that are already on the market. This argument was virtually doomed by the fact that almost every copyright extension act was applied retroactively. Grandfathering in existing copyrights is simply more administratively efficient than keeping track of "old" and "new" copyrights. The justices were thus faced with the possibility that accepting Eldred's argument might completely unravel U.S. copyright law. This issue of invalidating precedents partially obscures another, more fundamental one: the majority rejected the assumption that incentives in IP regimes represent a specific contract between the state and the copyright holder with a specific *quid pro quo*. Individual grants are justified by the *global success* of the copyright system, not by arrangements with individual copyright holders. Edward Samuel's *amicus* brief carries this logic all the way to the claim that the copyright clause does not even require the existence of a "public domain."[19]

In both *Donaldson* and *Eldred* those who felt their rights were being trampled were met with the same question: if these rights are so important, why hasn't anyone noticed them before?

Notes

1. Mark Rose, "The Author as Proprietor: Donaldson v. Becket and the Geneology of Modern Authorship," in Sherman and Strowel, *Of Authors and Origins,* chapter 2, 23–55.

2. Rose, *Authors and Owners,* 20.

3. Patterson, *Copyright in Historical Perspective,* 144.

4. Adrian Johns, *The Nature of the Book* (Chicago: University of Chicago Press, 1998), 188.

5. Rose, *Authors and Owners,* 23.

6. *Muret v. Nivelle,* 1586, quoted in Rose, *Authors and Owners* 20.

7. Joseph Addison, Tatler 101, Dec. 1, 1709; quoted in Rose, *Authors and Owners,* 36-37.

8. Daniel Defoe, Review, Feb. 2, 1710; quoted in Rose, *Authors and Owners,* 39.

9. Rose, "The Author as Proprietor," 25.

10. Rose, "The Author as Proprietor," 23.

11. All of the quotations below are drawn from Cobbett's *Parliamentary History of England,* volume 17, cols. 953-1003. I accessed this text from the website www.copyrighthistory.com/donaldson.html (July 25, 2003). This site is associated with the forthcoming book *The History of Copyright: A Critical Overview with Source Texts in Five Languages* by Karl-Erik Tallmo.

12. Christine MacLeod, *Inventing the Industrial Revolution,* 135.

13. Samuels *amicus* brief 2-3.

14. Dennis S. Karjala, "Judicial Review of Copyright Term Extension Legislation." *Loyola of Los Angeles Law Review* 36 (2002), 204n.

15. "Actually, Sonny wanted the term of copyright protection to last forever." Rep. Mary Bono 144 Cong. Rec. H9951 (daily ed. Oct. 7, 1998).

16. "If, on the other hand, Congress acted to extend every 50-year lease by the federal government for an additional 99 years at the government's current rent, there is no question that compensation would be required." Eben Moglen, *Eldred v. Ashcroft: Brief Amicus Curiae of the Free Software Foundation in Support of Petitioners.*

17. The *amicus* briefs opposing CTEA include an economist's brief, a constitutional law brief, an intellectual property law professor's brief, and a brief filed on behalf of the Free Software Foundation. For unparalleled analysis, see the Fall 2002 issue of the *Loyola of Los Angeles Law Review,* which is devoted entirely to the CTEA.

18. Appellant's opening brief, Appeal From The United States District Court For The District Of Columbia (D.D.C. No. 99-0065 [JLG]).

19. Samuels, *amicus* brief 6: "The constitutional phrase 'the progress of science and the user arts' is not directed primarily at the ultimate increase in the public domain."

5

The Natural History
of Intellectual Property

All societies create boundaries around specific forms of knowledge. The boundaries can be set by age, gender, membership in an elite, or a long process of education, but the ultimate purpose is the same. Knowledge is power, and the more tightly controlled the better, as the following example shows.

Prelude: The Philistine Blacksmiths

King Saul initiated his reign with a revolt against the Philistines. This revolt labored under many difficulties, not the least of which was the Philistines' technological superiority. Israel was still firmly in the Bronze Age. Bronze is expensive because it is an alloy of common copper and rare tin. By King Saul's time (approximately 1000 B.C.), the Philistines had mastered the art of making high-quality and durable iron tools. They were very happy to sell the *products* of Iron Age technology to the Israelites, but they were in no hurry to share the technology itself:

> There was not a single smith in the whole land of Israel, because the Philistines had reasoned: we must prevent the Hebrews from forging swords or spears. Hence all the Israelites were in the habit of going down to the Philistines to sharpen every ploughshare, axe, mattock or goad. The price was two-thirds of a shekel for ploughshares and axes, and one-third for sharpening mattocks and straightening goads. So it was that on the day of the battle of Michmash, no one in the whole army with Saul and Jonathan had either sword or spear in his hand, except, however, Saul and his son Jonathan [who presumably had bronze] (I Samuel 13:19-22, Jerusalem Bible).

There are several lessons to be learned from this example. First, the Philistines jealously protected the secrets of iron working by a *monopoly*. This monopoly served both political and economic ends. The Israelites were forced to arm themselves with prohibitively expensive bronze weapons or fight with agricultural tools. The fees charged were an efficient and unavoidable instrument of taxation. The monopoly also created an economic dependence on the Philistines that tended to keep the Israelites in a client role. Second, the Philistines were not hoarding anvils and bellows, but *knowledge*. Their advantage would only last as long as they were the only ones who knew the secret.

The first line of defense for valuable knowledge has always been to keep it a secret. Only those who become part of a group (families, clans, guilds, companies) are trusted with the dangerous knowledge. Unfortunately, history has shown again and again that all walls of secrecy eventually crumble. Even if no insider becomes a turncoat, human ingenuity eventually leads others to develop and master the "private" technology. Knowing that something is *possible* is often enough to start the search.

The Bible is understandably silent on whether or not the Philistines thought they owned their technology, or if their monopoly had a moral justification. Such niceties only become possible when there is some governing authority to grant and enforce rights, and when competitors are willing to cooperate. The Philistines were motivated by a shrewd sense of their self-interest, backed by the force of high-technology arms. Modern systems of intellectual property hide the threat of force behind a theory that justifies the monopolies of inventors in terms of the common good.

The ancient world revered literary works but disparaged inventors and artists. With the exception of Hero and Archimedes we know virtually nothing about creators of technology in the ancient world. The rewards of artistic creation were those of any other form of patronage: reputation and some degree of comfort in exchange for bolstering the patron's reputation and status. In the ancient world, artistic works were associated with those who ordered them rather than those who designed them (cf. the pyramids).

There are many reasons to create works, only some of them economic. People create works for prestige, to advance some project or ideology important to them, or to gain immortality through art, among other reasons. These reasons and others still operate today, in the academic world and in the new world of open-source software. To understand the dynamic of IP law we must understand how the development of printing led to a certain narrowing of focus: how works became commodified and assimilated into the economic world. We also need to examine the development and rationale of the first state-based system of IP: patents of monopoly.

Historical Background

Patents and copyrights originated in fifteenth- and sixteenth-century Europe, but were a culmination rather than a beginning. They united two ancient tools of state power (information control and economic monopoly) through a new moral rhetoric of legitimate rights.

State-granted Privileges

Monopolies were used throughout ancient history for patronage and to generate revenue for the state. Emperors and kings granted their favorites monopolies on some necessity of life, often for a share of the proceeds. The monopolist would then enforce the monopoly himself and do his own record keeping. However, as a form of taxation, such a system invites corruption: monarchs rarely earned much from private monopolies. As a way of bestowing largesse, it was a triumph: the monarch could give someone else's wealth to a favorite without increasing the cost of government. In effect, monopolies externalized the cost of royal largesse. In an age of monarchs, no justification beyond royal prerogative was expected.

The modern system of IP law began in fifteenth-century Europe.[1] Monopolies were nothing new, as the example of the Philistines shows. What was new was the fact that "letters patent" were beginning to be justified in terms of social good and moral right. This justification did not always correspond well with reality, as we shall see.

The first patent in the modern sense was granted in 1467 in Berne, for the manufacture of paper. In 1469 Johann von Speyer was given the exclusive right to practice printing in Venice for five years.[2] In 1507 the Venetian council of ten granted a twenty-year patent for a secret process of mirror-making. This same industry was established in France with the grant of a ten-year monopoly in 1551. The rationale for these grants was probably what we would now call industrial policy: an attempt to foster growth of new technologies in a particular country.

The system of letters patent was developed most fully in Tudor England.[3] England's geographical isolation made enforcing patents much easier than it was on the continent, and the system of patronage created by patents was very attractive to a relatively weak central government with limited funds.

The original intention of the patent system was to encourage innovation. Francis Bacon offered the following definition in a speech to Parliament on November 20, 1601: "If any man out of his own wit, industry, or endeavor, find out anything beneficial for the commonwealth, or bring any new invention, which every subject of this realm may use; yet in regard to his pains, travail, and charge therein, her Majesty is pleased (perhaps) to grant him a privilege to use the same only to himself, or his deputies, for a certain time: this is one kind of monopoly."[4]

The patents of monopoly proved to be an economic and political disaster for England. The system quickly degenerated: "Originally, the patents had been given for ten years, but by this time [1570-1580], twenty, twenty-one, and thirty years had become more common terms, the practice of reissuing had commenced, and patents were no longer confined to new arts."[5] Patents were granted on almost anything of economic value, including salt and soap. Patents were also granted on whole classes of goods and services (e.g., playing cards, rag-collecting, printing Latin grammars). Patentees had private police forces that could harass competitors with virtual impunity. Since the granting of patents was a purely a matter of royal prerogative, only those with financial or personal influence on courtiers could receive them.

Patents of monopoly became enormously unpopular, and several attempts were made to reform the system. Elizabeth's last speech was an apology to her subjects for the abuses of the patent system.[6] James I created a permanent commission to investigate requests for patents that unfortunately proved to be both inefficient and corrupt. Following Elizabeth's example, James' final legislative act was the Statute Of Monopolies of 1624. This act repealed all monopolies *except* letters patent that met the following conditions:

> Provided nevertheless, and be it declared and enacted that any declaration before mentioned shall not extend to any letters patents, and grants of privilege, for the term of one and twenty years or under, heretofore made *of the sole working or making of any new manufacture within this realm, to the first and true inventor or inventors of such manufactures which others at the time of making of such letters patent and grants did not use*, so they be not contrary to the law or mischievous to the state, by raising of the prices of commodities at home, or hurt of trade, or generally inconvenient, but that the same shall be of such force as they were or should be if this act were not made, and of none other [emphasis added].[7]

The statue limits new letters patents to "first and true" inventors and to a term of fourteen years. The act also clearly states that patents are fully under the regulation of common law, explicitly making the courts the arena for disputing patents and monopolies.

Copyright: Property Rights as Social Control

The development of the printing press played a key role in the emergence of intellectual property theory. Printing had profound effects on history and culture which are still being sorted out.[8] The development of the printing press completely transformed the economics of publishing. The scriptorium has a fixed cost and rate of production: the amount of time one copyist could copy a manuscript. The volume of output could be increased by added more copyists, but not the rate. With a printing press, there was a relatively expensive stage of typesetting, followed by a very inexpensive and very rapid stage of production. A team of two good pressmen were expected to print 250 pages an hour, 1200 to 1500 pages in a typical 14-hour day.[9] No copyist could average 600 pages a day!

The development of printing also changed the social dynamics of reading and writing. Printers required capital, skilled workers, and expensive equipment. Their financial requirements and centralization made them more vulnerable to the power of kings than any group of scribes would have been. In sixteenth-century England, monarchs feared the power of the printers to publish "heretical schismatical blasphemous seditious and treasonable Bookes, Pamphlets and Papers." Some form of control was clearly necessary, and in fact no fewer than three overlapping systems developed for regulating printing.

The earliest system was simply patents of monopoly. Monopolies were granted for publishing prayer books, Latin grammars, and ABCs. Such books presumably had a constant market that would make investing in a patent worthwhile. From Henry VIII through the restoration, the government also imposed various regimes of censorship and licensing books. However, the most enduring system was that of the Stationer's Company Monopoly.[10]

In 1557 Queen Mary granted a charter to the Stationer's Guild (or "company") that granted the Stationer's a monopoly on "the art or mistery of printing." The rationale for this grant was stated in the preamble to the charter:

> Know ye that we, considering and manifestly perceiving that certain seditious and heretical books rhymes and treatises are daily published and printed by divers scandalous malicious schismatical and heretical persons, not only moving our subjects and lieges to sedition and disobedience against us, our crown and dignity, but also to renew and move very great and detestable heresies against the faith and sound doctrine of the Holy Mother Church, and wishing to provide a suitable remedy in this behalf. [11]

The right to print "copy" was restricted to members of the Stationer's Company, who could only print works listed in the Stationer's register. These "copy rights" were the permanent property of their registered owners. All disputes about copyrights were settled by the Stationer's Company rather than the courts. The charter gave members of the Stationers the authority to search out and destroy unauthorized books, making them the chief enforcers of censorship.

The Stationer's Company acted under this charter from 1557 until 1710, when the Statute of Anne ended their monopoly on printing. Several details of the Stationer's monopoly are important in what follows:

- Copyright was a right of publishers, not authors. We shall have a great deal more to say about this in chapter 10.
- Copyright applied to *all* books, not just "original works." As originally formulated, copyright represents a monopoly on making money from publishing, rather than any reward for creativity.
- Copyrights were perpetual rights of ownership, like owning a house or a piece of land. Once granted, they could never be infringed by anyone without action by the Stationers.
- The Stationer's Company formed a closed and private system of ownership that operated outside the legal system. It was in effect a private system of law protected by royal privilege.
- Copyright's ultimate rationale was politics, not equity. In the words of L. Ray Patterson, "The government was not really interested in copyright as property, only as an instrument of censorship . . . As a device

to control the distribution of printed material, copyright was ideal, since it combined so well the interest of the government with the self-interest of the copyright owners."[12]

The Stationer's Company managed to hang onto its copyright through the English revolution and the Restoration by demonstrating its zeal in carrying out the censorship policies of whatever party was in power. However, censorship became less and less popular: by 1662, licensing acts were limited to two years at a time, and were finally abandoned in 1694, when the House of Commons refused to renew it at all. The Commons gave eighteen reasons for refusing to renew the act, including the following:

- By requiring registration with the Stationer's Company, the Stationers "are impowered to hinder the printing all innocent and useful Books: and have an Opportunity to enter a Title to themselves, and their Friends, for what belongs to, and is the Labour and Right, of others."
- The act allowed the Stationers to monopolize the publication of all books, including classics. Furthermore, the act "prohibits the importing of any such Books from beyond the Sea: whereby the Scholars in this Kingdom are forced not only to buy them at the extravagant Price they demand, but must be content with their ill and incorrect Editions."
- The act prohibited the printing and importing of all offensive books, but never defined what made a book "offensive."
- The penalties of the act were harsh and inconsistent with its own stated goals. There were no penalties for publishing treasonable and seditious books, but "there are great and grievous Penalties imposed by that Act for Matters wherein neither Church or State is concerned."
- "Lastly, There is a Proviso in that Act for John Streater, That he may print what he pleases, as if the Act had never been made; when the Commons see no Cause to distinguish him from the rest of the Subjects of England."[13]

The reformers of copyright were faced with the same problem as those who had reformed the patent system in 1624: to continue the system but to make it something more than the arbitrary exercise of royal privilege. The first step in this rationalization would be to insist on some non-arbitrary relationship between the holder of the rights and the protected property. Two parties seemed likely candidates: publishers or authors. In 1624 the patent system had begun to move in the direction of authors: the statute restricted patents to the "first and true inventor or inventors of such manufactures which others at the time of making of such letters patent and grants did not use."

Developments since the Eighteenth Century

The last two centuries have seen many changes in the landscape of intellectual property. Three are particularly important for what follows: the emergence of IP protection for visual art, the development of recording technology, and the development of what the Russians called "samdizat" (self-publishing) technology.

Congress amended the U.S. copyright law in 1865 to cover photographs, though *which* photographs were covered was not addressed by the Supreme Court until 1885.[14] In 1903 the Supreme Court extended copyright protection to advertisements. The extension of copyright to photographs and advertisements involved crossing some significant boundaries. Pictures are not words, the traditional domain of copyright protection. Furthermore, photographs were not seen by all as involving creativity. In *Sarony v. Burrow-Giles*, the defendant argued that his unauthorized reproduction of a photograph was not copyright infringement because "A photograph being a reproduction on paper of the exact features of some object or some person, is not a writing of which the producer is the author."[15] This argument was rejected in a unanimous decision by the Supreme Court in 1885, though only on the grounds that the photograph in question (a portrait of Oscar Wilde) was in fact, a product of artistic effort.

The requirement that a work be "artistic" was removed by Supreme Court in their decision on *Bleistein v. Donaldson Lithographing Company* in 1903. Writing for the majority, Oliver Wendell Holmes Jr. argued "A picture is none the less a picture and none the less a subject for copyright that it is used for an advertisement."[16] This decision indicated a shift in intellectual property law from a narrow focus on the intent of a work ("Does it advance knowledge?") to a broader protection based on the medium of a work ("Is it a picture?").

Another major change was driven by the development of recording technology. The first real "recording" technology came with the development of player pianos. Is a player piano roll a "copy" of a musical composition, or a completely new kind of work? The problem became even more acute with the development of the phonograph and motion pictures, and variations of this problem recur with development of each new medium. Consider the question of whether or not radio stations should be allowed to play recorded music.[17] Congress had granted authors the right to collect royalties for the public performance of their works in 1897, but is the act of putting a record on a turntable a "performance"? Though now largely forgotten, performance rights dominated copyright law for nearly fifty years.

The final change we will consider here is the development of copying technologies so inexpensive and simple that they could be used by individual consumers. Xerox machines, cassette recorders, VCRs, and CD burners make it possible for almost anyone to make good copies of a work at very low cost. The development of digital media can only accelerate this trend.

Technology and the Convergence of Author, Publisher, and User Rights

We can get a good sense of the classical understanding of intellectual property by looking at the IP situation in the United States around 1830. Both copyrights and patents were relatively rare and closely tied to perceived social value. Most written works never even aspired to copyright. No protection was recognized for non-American works.

The expansion of copyright protection to photographs, commercial art, piano rolls, sound recordings, motion pictures, radio and television programming, and computer programs also reflected a change in the legal rhetoric surrounding copyright. Even as the romantic figure of the author received more and more attention, the legal reality of copyright was shifting from perceptions of artistic and social value to a purely "commodified" economic framework. This transition is crucial for understanding the current IP crisis, since the erosion of analysis of value underlies the erosion of social utility as a relevant criterion for granting intellectual property rights. That erosion has become so complete that it is hard to even imagine what a socially based IP regime would now look like.

The process of commodification has also affected patent law, which has expanded into a variety of new areas in the last fifty years. It reaches its ultimate expression in the right of publicity, the doctrine that people own distinctive features of their appearance and behavior.

The development of technology to record performances was followed by the emergence of broadcast media. Unlike traditional publishers, radio stations offered, as a *service*, an endless stream of music and performances created by others. But what gave them the right to profit from the efforts of others? On the other hand, what gave artists and publishers the right to restrict the *use* of their works if no *copying* was going on? The legal issues associated with performance media began with player piano rolls and are still being fought today. Here we will examine one possible solution: the practice of compulsory licensing.

The term "expansion," while accurate, should not mislead the reader into assuming that the process was orderly, steady, or particularly peaceful. Technological progress has done considerably more than simply introduce new forms of expression that need protection. The emergence of recording media has fundamentally changed the relationships between authors and users on the one hand, and publishers on the other. The proliferation of "personal" copying technology since the 1960s has made piracy a do-it-yourself option for users in a way that was never possible before. The emergence of digital media realizes a publisher's worst nightmare: the ability of pirates to make an unlimited number of *perfect* copies for almost no marginal cost.

Digital media are a byproduct of the almost total computerization of technology in the last fifty years. Understanding computer programs as property is an important issue that we will be considering shortly. Here I want to draw attention to another consequence of computerization: the emergence of what I call *active media*. Publishers are beginning to wrap software around their products that can control the use of those products. It is as if each CD or video tape contained a tiny lawyer who was always there to enforce the publisher's IP rights. The enormous flexibility of software means that publishers can introduce almost any level of control they desire into their products. They no longer have to depend on the good will of users or beg for their rights to be observed. Most importantly, active media removes the need for publishers to rely on the government to coercively enforce their rights: they can implement any protections or restrictions they choose.[18] In a fully digitized world, IP law would be irrelevant: it would be replaced by arbitrary contracts between publishers and users. This would also be the end of any understanding of IP as some kind of social contract designed to promote social utility, or of the idea that users somehow have *rights* that are independent of terms that publishers wish to dictate to them. Active media replaces the publisher's exclusive right to *copy* with an exclusive right to control *any use whatever*.

Photographs

The extension of copyright protection to sheet music in 1831 was relatively uncontroversial.[19] The extension of copyright to photographs in 1865 was significantly more controversial. The issue came squarely down to the sense in which a photograph was an expression of creativity. Wasn't a photograph a mechanical reproduction of the appearance of a scene? How could the mere capture of light from a scene be considered original or creative?

It is worth reflecting for a moment on the difference between a piece of sheet music and a recorded performance of the same music. Sheet music represents a piece of music using a language designed to capture the distinctive features of music (tone, rhythm, dynamics, etc.) in a form that can be used to recreate the pattern of sounds that *is* the music. This notation represents the music as a collection of ideal units (significantly called "notes") that are realized through performance into sound waves. Now consider a recording of the same piece of music. The recording captures not the *music*, but a *performance* of the music as a collection of sound waves. The recording can be used to reproduce the sound waves, but what it reproduces is the recorded *performance*. The relationship between sheet music and performance is type-token, while the relationship between a recording and a performance is, as it were, token-token. This is reflected in the fact that a recording represents a collection of sound waves, not instructions about how to recreate the piece. The "language" used to represent a piece of music on magnetic tape or on a CD cannot distinguish between music, speech, natural sounds, or anything else. That language is designed to be processed by a machine that can reproduce patterns of sound waves, and which does not require any understanding of what music or sound are. This, of course, is why it doesn't take years of training to "make" music with a tape player.

Is this difference in levels relevant in deciding whether or not recordings should be considered works? At first it seemed decisive. The argument went like this: when a musician performs, she creates sound waves that fill up the performance space and are available to anyone in the audience. A person who uses a tape recorder to "capture" those sounds is not doing anything different than a person who takes their supper home from a restaurant in a doggie bag.[20] Selmer Bringsjord uses examples like this to argue that there is no principled difference between "copying" and "perceiving" something.[21] Nor is anyone who copies a sound recording, since *the recording* isn't a "creative work" on the part of the artist or the person running the recording device.

This second line of argument was first considered in 1885, when the 1865 statute making photographs subject to copyright was challenged by the Burrow-Giles Lithographic Company.[22] Burrow-Giles directly challenged the copyrightability of photographs by arguing that "A photograph, being a reproduction on paper of the exact features of some natural object or of some person, is not a writing of which the producer is an author."[23] In a unanimous decision, the court ruled that Sarony's *composition* and *active physical arrangement* of his photograph was "giving visible form" to "his own original mental conception." The choice of lighting, props, and film stock was interpreted as directly analogous to an author's choice of word and narrative style: thus a photograph embodies the idea/expression distinction. The fact that the photograph captures features of the public world is no more relevant than a painter's use of a model deprives his painting of originality.

Sound recordings became an issue with the development of the player piano and the phonograph.[24] Revisions to the copyright law in 1909 specifically prohibited the unauthorized mechanical reproduction of musical compositions. The 1909 act is also noteworthy for introducing the concept of *compulsory licensing*, which was to prove a useful tool in resolving issues associated with radio and television broadcasts.

Compulsory Licensing: An Institutional Response to Anticommons Problems

Granting copyright protection to recordings and performances creates transaction–cost problems for both publishers and users. Granting a publisher the right to payment of royalties is meaningless if collecting them is impossible or too expensive (there is no point in spending ten dollars to collect five). Users face the familiar problems associated with the anticommons if there are many owners to be compensated for copying works.

Copyright protection for recordings and performances created *different* transaction-cost problems for publishers and authors. Publishers face the costs associated with finding and negotiating with many individual authors. Authors face the problem that they have relatively little leverage as individuals: the costs of enforcing their rights through the legal system can be prohibitively expensive.

Compulsory licensing attempts to solve these problems by simply eliminating the element of negotiation. The rights of authors shift from being "property" rights (the right to refuse access) to being "liability" rights (the right to be compensated after use).[25] Composers were paid a fee (imposed by statute rather than by market mechanisms) for each reproduction of their composition by a recording.

Compulsory licensing creates a mechanism whereby new types of reproduction can be introduced without a potentially endless process of negotiation. But it has much broader implications as well:

- Compulsory licensing *replaces* the paradigmatic IP right of owners (the right to control or forbid reproduction) with another right (the right to be compensated). In this respect, compulsory licensing is similar to eminent domain: the state overrides individual property rights in the interest of social utility.[26] As we shall see later with pharmaceutical patents, the line between compulsory licensing and expropriation can be a very fine one indeed.
- Compulsory licensing strongly favors users, since it eliminates the monopoly power traditional copyright gives authors and publishers. The payment to owners by publishers is fixed by statute, not by negotiation; and since publishers are not granted monopoly rights, they can't simply raise the price of copies arbitrarily.

Compulsory licensing of copyright created an environment in which a market for recordings could successfully develop. However, it left the issue of using recordings for profit (by broadcasting them) unresolved. Issues associated with performance and broadcasting were finally resolved by the emergence of collective but *private* licensing authorities such as ASCAP and BMI.

Performances

Certain types of work can only be realized through their performance, most notably drama and music. But is public performance a form of copying? If it is, how is performance covered by copyright? American copyright law was amended in 1856 to grant an author exclusive performance rights for dramatic works, and in 1897 to grant exclusive performance rights for musical works.[27]

French copyright law was forced to deal with the issue of performance much earlier.[28] After the revolution there was a protracted debate about whether or not the *Comedie Francaise* and other licensed theaters should retain their monopoly on performance of dramatic works (a privilege left intact by the new revision of copyright law). The playwrights agitating for repeal of the privilege asked for an exclusive performance right lasting until five years after the death of the author. Unlike their aggrieved British brethren, they presented themselves as "contributers to 'public property' and guardians of the nation's cultural commons."[29] The rights were readily granted, perhaps because of the importance of theater as a tool for public education and propaganda. Objections from theater owners centered around the fact that most dramas were works-for-hire: there was no question that they were in some sense *literary* works.

The issues associated with recording performances were made even more difficult by the emergence of radio broadcasting as a business in the 1920s. Was the broadcast of a musical performance (or worse yet, of a *recording* of a performance!) a "public performance"? If it was, then the 1897 law meant that radio stations would have to negotiate agreements with all of the composers whose work they played on the radio.

Fixation

The court's argument in *Sarony* also opened the way for a fundamental change of metaphor in copyright law: replacing the concept of *writing* with the much more general idea of *fixating*. According to the 1976 Copyright Act, the object of copyright law "subsists, in accordance with this title, in original works of authorship fixed in *any tangible medium of expression, now known or later developed, from which they can be perceived, reproduced, or otherwise communicated, either directly or with the aid of a device.*"[30] Virtually any physical object or event can be the fixation of a work as long as it can be interpreted as a "medium of expression."

Active Works

Copyright protection was originally applied only to very specific types of physical objects (words on paper and maps). The shift of focus from writing to fixating made virtually any type of object or event potentially subject to copyright. Yet even after broadening the nature of a work so tremendously, works remained essentially inert. They are the cargo or freight that gets moved around in broadcast media. Even calling them "cars on the information superhighway"[31] is misleading: cars, after all, are "auto" mobiles. In this section we will consider active media: expressions that interact dynamically with their environment. The paradigmatic active medium is computer software.

Computer Software

Computer software was not *specifically* covered by copyright law until 1980, when an amendment to the 1976 copyright act specifically designated some special restrictions on copyrighting computer programs (defined as "a set of statements or instructions to be used directly or indirectly in a computer in order to bring about a specific result").[32]

Figuring out exactly *what* software is (or should be considered to be) is rather harder than the definition above might suggest.[33] Something "intended to bring about a specific result" sounds more like a machine than a literary work. And indeed, computer software was eventually made patentable. However, there are a few analogies that are worth examining:

- Like literary works, computer software is written using organized systems of symbols. These symbols generally do not dictate the exact form of a program, since there are usually many potential programs that can perform a given function. Considerable skill is required to create a program that efficiently meets several competing design criteria. There is also some room for the expression of originality in coding software. Thus the creation of software is a process that will sustain an analogy to the process of literary creation.
- The stored–program design of most computers means that a single piece of hardware is capable of hosting many different pieces of software. It is also perfectly possible for many different computers to host "tokens" of the same program "type." Thus, a computer program shares some of the immateriality we have already seen in literary works.
- The practice of writing programs has developed as a distinct intellectual culture. There is an emerging "literary heritage" of technique and aesthetics that functions in many of the same ways that literary criticism functions for literature.

There are also powerful pragmatic reasons for granting IP protection to software. Creating good software requires a large investment of time, energy, and money. At the same time, programs (as executable files) are almost effortlessly copyable. And unlike "the great American novel," software is eminently useful. There are almost no limits on what a computer system can do if it is loaded with the right kind of software. The utility of a market for software is obvious. Without some form of IP protection, it is very difficult to see how it would even be *possible* for a market in software to exist.

The question then became: what kind of IP protection is appropriate for computer software? It is a witness to the protean nature of software that the ultimate answer ended up being "almost all kinds." As noted above, programs became copyrightable in 1980. In the 1983 case *Diamond v. Diehr*, the Supreme Court ruled that a computer program was patentable subject matter.[34] If a company makes reasonable efforts to keep the source code of a program secret, it can also be protected as a trade secret through such mechanisms as non-disclosure agreements. There is thus no shortage of legal tools that a software developer can use to protect his programs. In most circumstances, software developers rely on trade secrets, but almost all of the commercial software I've worked on now contains a copyright notice as well. Patents give very strong protection, but are hard to get: "The level of creativity required to get a patent is too high, and most computer programs don't qualify. In any event, it takes about a year or two to get a patent, and in the software business, one year and you're obsolete."[35]

This arsenal of legal tools almost immediately bumped into a profound problem created by the nature of computer software: an executable program is just a data file like any other. The functioning of a computer requires the system's ability to read and copy files, and in particular the installation of new software requires the ability to copy programs from one medium to another. Thus if I buy a piece of software, nothing but the law prevents me from letting all of my friends copy it too.[36] This has not been good enough for software makers, who have used the active nature of software to take the law into their own hands.

Copy Protection Schemes

The first technical attempts to defeat copying involved "copy protection": the installation disks were written in various quirky ways that made it possible to read them but not copy them. This approach fell from favor very quickly for a variety of reasons:

- Early personal computers were so unreliable that the lifespan of install disks (and any other disks, for that matter) was extremely short. The ability to back up files is not optional, especially for businesses. Customers resented the vulnerability that copy-protection imposed on them. In addition, the 1980 amendment that legitimized software copyright exempted "archival" or backup copies from copyright. This meant that software manufacturers had no legal right to prevent users from backing up their software.
- Computer users were often very technically gifted and highly motivated to defeat copy-protection. No "secret quirk" of operating systems or disk formats remained secret very long.
- "Copy busting" software, which was perfectly legal, quickly became available for the less sophisticated users. (Thanks to the passage of the Digital Millennium Copyright Act, developing and disseminating such software is now a felony punishable by five years in federal prison and a $500,000 fine.)

Within a couple of years, software manufacturers had retreated to schemes involving the "activation" of software by entering some key string. When well administered, this provided some control of copying, but remained extremely easy for users to circumvent. More indirect approaches were also tried: software packages began to "bloat," with sheer size discouraging copying. The most successful approach (though only for Microsoft) was a shift back to "bundling": your computer came with software pre-installed, and the cost of the software was buried in the cost of the PC. People who receive a PC loaded with software have few incentives to make illegal copies.[37]

Microsoft's latest version of the Windows operating system (Windows XP) takes a much more aggressive approach to controlling installation. At Install time, the system generates a "starter" key value. To get the complete key (and thus activate the software) the user has to either connect to a Microsoft website or call a toll-free phone number and give Microsoft the key. Microsoft then uses their own secret software to generate the actual activation key. Thus Microsoft can explicitly track whether or not a particular copy of the installation software has been used before.

The enabling technology of the new wave in copy protection is strong encryption.[38] Installation media store their data in an encrypted form that is virtually impossible to decrypt without a key that is only available from the manufacturer. The next step in control is to skip decryption entirely, and require the presence of a program that decrypts and loads the program whenever certain conditions are met. The current term for such programs is "digital rights managers." [39]

Trusted Systems and Digital Rights Management

The joy of getting something for nothing has led some to the belief that technology is on their side, that technology inevitably will bring about a completely open world. They're wrong. The last half-century of IP theory can be seen as a race between law and technology, with technology favoring users over authors. But the technology that supports limitless perfect copies can also be a peerless tool of control. Imagine a book you could only read by sticking your credit card into it, or that couldn't be lent to anyone (no one else can open it), or that disintegrates after being read three times. Imagine that the only way to read a book was by giving your name, address, and phone number to the publisher, who can then do anything they want with the information. Imagine that the same book cost you $20.00 but cost the library $5000.00. How many books would the library have?

All of these wonders (or horrors) have become possible through the insight that copy protection technologies can be used to control access to any other kind of digital media. As long as all of the hardware involved (host computer, user's computer, printer) were designed to automatically enforce access controls, then publishers could control *all* potential uses of a work. All of the possibilities listed above (and more) are possible.

Trusted systems cross the line between controlling copying of a work and controlling the *use* of a work. Until now, the independent physical existence of copies meant that authors and publishers lost control of them once they left the bookstore. This has exposed publishers to the risk of loss from unauthorized copying, but it has also made many socially useful institutions possible, such as libraries, used bookstores, and buying books at a garage sale.[40] Copyright is, after all, only intended to control *copying*: it is hard to see what would give the publishers of digital works the right to control all possible uses of their work.

In his article "Shifting the Possible: How Trusted Systems and Digital Property Rights Challenge Us to Rethink Digital Publishing,"[41] Mark Stefik argues that trusted systems actually increase access to digital media by making it possible to put a price on any possible use of a work. You could buy a "one-time-use" copy of a work for less than buying a permanent copy: libraries could have a limited number of "free" copies of a popular work and charge for immediate access to a non-free one. Expiration of copyright would mean that you could either download a free copy from the Library of Congress or buy a *still-protected* copy from the publisher.

While Stefik notes in passing that "Trusted systems shift the balance and put more power in the hands of publishers,"[42] he seems remarkably unperturbed by this fact. He sees this as an organizational problem and proposes an organizational solution. Issues about fair use and general social policy would be handled by an institution he calls the "Digital Property Trust," which "would have as its objective *the promotion of commerce* in digital works" (emphasis added). Those who feel they need fair use of works could apply to this organization for a license, which would be granted after the applicant (supplicant?) could "demonstrate an understanding of the principles of fair use and copyright." It would be worth it: "These licenses would have associated special discounts or free use for certain kinds of works and *perhaps* fewer limitation on rights." The DPT would also provide insurance against the possible misbehavior of licensees: it is not too hard to guess who would end up paying for it.

Stefik's proposed solution assumes that since trusted systems allow complete control over the use of a work, and since the work is owned by the publisher, the publisher has the right to complete control over their work. But the fact that a technology allows me to control something forever doesn't mean I therefore have the *right* to impose that kind of control. The default assumption in commerce is that sale of something involves completely alienating to the seller. Why should IP be different? Stefik rehearses the familiar argument for why publishers must have control of digital works, but this only justifies control of *copying*, not all the other controls he proposes. Even if trusted systems would make all kinds of use possible, it would still abolish many rights that already exist, and that people can exercise without paying money. Stefik seems to believe that the term "right" means something like "privilege": my complete sovereignty over my property means that the only rights you have are ones I choose to grant to you. But even if we ignore the question-begging nature of that argument, interpreting rights purely as grants fails to take adequate account of their moral urgency.

Even passive rights involve an *obligation* toward the recipient, and active rights go much farther. As Richard Tuck puts it, "According to this view, to have a right to something is more than to be in a position where one's expressed or understood want is the occasion for the operation of a duty imposed upon someone else: it is actually in some way to impose that duty upon them, and determine how they ought to act towards the possessor of the right."[43]

Non-public Media and Contractual Rights

In the 1984 *Sony Corp. of America v. Universal City Studios* case, the Supreme Court ruled that the common practice of using a VCR to "time-shift" programs was not an infringement of copyright.[44] According to Jane Ginsburg, "The court held that because the public had been 'invited to witness . . .[the programs in their] entirety free of charge,' copying them for time–shifting purposes was 'fair use' of the copyrighted works."[45] The copying was legitimate because it was just a means to exercise the invitation for free viewing implicit in a public broadcast.

Some legal theorists have wondered if there is an unrestricted right to read that is the mirror-image of the owner's right to control copying.[46] Something like that seems to be at work when you turn on a radio or check a book out of the library. One might even think that such a right is implied by the statutory claim that the goal of the IP system is "to promote the advancement of science and the useful arts." Isn't the point of IP production to increase the supply of works? And what is the point of increasing the supply unless the ideas are made available somehow?

This seems reasonable. But what exactly would such a right look like? We regularly restrict people's access to various kinds of information in the interests of privacy. We also put *economic* restrictions on the right to access works. I have a right to see a movie only if I buy a ticket. Convenience store clerks frequently remind people "this isn't a library." If you want to read, you have to pay first. Cable TV and Internet access are only available to those who pay to hook up to the service.

Cable companies and Internet service providers aren't theaters and bookstores: their primary product is not content, but the right to access content. Since no one has to get cable, cable operators can argue that they are not providing any kind of public service: they are free to set any terms that the market and regulators will bear for their services. When accessing a content provider, the social contract of IP law is replaced by the arbitrary contract between customer and seller.

Before looking at the ethical consequences of the argument above, it is worthwhile to reflect briefly on the role that "economics and physics" have played in defining copyright. Several physical and historical contingencies have shaped the most basic features of copyright:

- Making books (and especially making copies of books) was originally a labor and capital-intensive process. There was no such thing as a "home printing press." The only likely motive for copying without permission was piracy. Controlling printing has been the preferred method of controlling the press in most dictatorships.[47]
- It is the nature of printing that startup costs are much greater than production costs. Most of the cost of traditional printing is associated with compositing the printing plates, not actually printing the copies. Therefore institutional rules have tended to treat publishers as if they were investors.
- Because a book is a discrete physical thing, a publisher naturally loses control over it after it is sold. This is the practical basis of the first-sale doctrine.
- Because a book is an enduring physical thing, there is no way for a copyright owner to impose time limits on user's rights.
- Printing can only capture texts, not performances. A musician who wanted to maintain exclusive use of his compositions had no choice other than to not publish them.
- Because a copy of a book is like any other personal possession, strictly controlling them requires a very large-scale intrusions into the lives of users.
- Books can only be used by one person at a time. I can't use my copy of the book while you have it (a fact that takes some of the sting out of the loss of control inherent in the first-use doctrine).
- Before recording technology, performances were inherently ephemeral. There was no way for either an owner or a pirate to copy them and then vend them.

The constraints created by the physical nature of books have determined many of the boundaries between owner's rights and user rights. But if that physical nature changes, do the rights change? Is it possible that user rights were simply a historical accident created by the imperfect nature of books?

IP Rights as Contractual, Statutory, and Natural

The *NII White Paper* is as clear about the nature of fair use as it is about most IP matters: fair use in no sense reflects any kind of natural right of access to works. "Although sometimes referred to as 'rights' of the users of copyrighted works, ' fair use' and other exemptions from infringement liability are actually limitations on the rights of copyright owners. Thus, as a technical matter, users are not granted affirmative 'rights' under the Copyright Act; rather, copyright owner's rights are limited by exempting certain uses from liability."[48] Copyright owners must be vigilant not to allow the government to pile insult upon this existing injury:

> Some participants [at an NII working group conference on fair use] have suggested that the United States is being divided into a nation of information "have" and "have nots," and that this should be ameliorated by insuring that the fair use defense is broadly generous in the NII context. The Working Group rejects the notion that copyright owners should be taxed–apart from all others–to facilitate the legitimate goal of "universal access." [footnote: The laws of economics and physics protect producers and equipment and tangible supplies to a greater extent than copyright owners. A university, for example, has little choice but to pay to acquire photocopy equipment, computers, paper and diskettes. It may, however, seek subsidization from copyright owners by arguing that its copying and distribution of their works should, as a fair use, not be compensated.][49]

An obligation to allow some non-economic uses of copyrighted works is here seen as a discriminatory tax on authors. The footnote suggests that the reason that copyright owners might be singled out is that they lack the natural protection provided by physics and economics.[50]

Both the *NII White Paper* authors and their critics *might* agree that the fair use issue is an ontological problem. It is undoubtedly harder to maintain exclusive control of works than of paper and diskettes. But it seems just as reasonable to infer that asserting ownership rights over non-physical things is "an exercise in futility"[51] as it is to bemoan the "discriminatory" treatment of authors. If society won't enforce the full spectrum of rights that are being claimed by authors, then perhaps the problem is in the claims rather than in the users.

The technology of trusted systems promises to give authors a way out of such social demands. With the technology of trusted systems (and with the legal help of the Digital Millennium Copyright Act and the UCITA additions to the uniform commercial code), authors may be able to enforce their rights without waiting for the government. As we have seen, trusted systems allow authors to specify and *automatically enforce* any terms of *use* they choose. Works will *never* pass out of the author's control, undermining principles like the right of first sale or fair use. Death and transfer of ownership will invite chaos, as users suddenly discover that their rights with respect to a particular work have suddenly changed without their knowledge or consent. There will be no social contract: only private property protected by perfect technological barriers. The only "rights" will be private property rights.

Notes

1. Price, *English Patents of Monopoly*, 3.

2. Price, *English Patents of Monopoly*, 3.

3. Price, *English Patents of Monopoly*, 5.

4. Price, *English Patents of Monopoly*, 154.   In the same speech Bacon also advocates monopolies as a way of stabilizing commodity prices.

5. Price, *English Patents of Monopoly*, 8.

6. Price, *English Patents of Monopoly*, appendix.

7. Price, *English Patents of Monopoly*, 137–138.

8. Johns, *The Nature of the Book*, 90.

9. Johns, *The Nature of the Book*, 93.

10. Johns, *The Nature of the Book*, 187–248; Patterson, *Copyright: A Law of User Rights*, 20–27.

11. Rose, *Authors and Owners*, 12.

12. Patterson, *Copyright: A Law of User Rights*, 26.

13. See "John Streater and the Knights of the Galaxy" in Johns, *Nature of the Book*, chapter 4.

14. Goldstein, *Copyright's Highway*, 58–63.

15. Goldstein, *Copyright's Highway*, 59.

16. Goldstein, *Copyright's Highway*, 61.

17. Goldstein, *Copyright's Highway*, 66–75.

18. The role of government would be to coerce manufacturers to implement the infrastructure of active media, and to criminalize circumvention of the technology (or even *discussion* of the technology). Government protects IP rights by refusing to allow anyone to "opt out" of the system for any reason.

19. Samuels, *Illustrated History of Copyright*, 32.

20. There are even people (like Mozart) who are capable of hearing a piece of music once and then writing down a perfect transcription of it.

21. Selmer Bringsjord, "In Defense of Copying," *Public Affairs Quarterly*, 3, no.1 (1989): 1–9.

22. Goldstein, *Copyright's Highway*, 58–60.

23. *Burrow–Giles Lithographic Company v. Sarony* 111 U.S. 53 (1884). Quoted in Goldstein, *Copyright's Highway*, 59.

24. Goldstein, *Copyright's Highway*, 64–67.

25. Robert Merges, "Institutions for Intellectual Property Transactions: The Case of Patent Pools." Unpublished working paper accessed at the author's website www.law.berkeley.edu/institutes/bclt/pubs/merges/pools.pdf 12–16.

26. "Overrides" is not really the right term here, since some would argue that IP rights are created by the state in the first place.

27. Goldstein, *Copyright's Highway*, 67–69.

28. Carla Hesse, *Publishing and Cultural Politics in Revolutionary Paris, 1789–1810.* (Los Angeles: University of California Press, 1991), 113–116.

29. Hesse, *Publishing and Cultural Politics*, 117.

30. Emphasis added. Quoted in Samuels, *Illustrated History of Copyright*, 127.

31. Jane Ginsburg, "Putting Cars on the 'Information Superhighway': Authors, Exploiters and Copyright in Cyberspace," *Columbia Law Review*, 95 (1995): 1466.

32. Samuels, *Illustrated History of Copyright*, 82.

33. Peter Suber, "What is Software?" *Journal of Speculative Philosophy*, 2 (1988): 89–119.

34. Samuels, *Illustrated History of Copyright*, 80-81.

35. Samuels, *Illustrated History of Copyright*, 80.

36. A famous joke: "What's the projected market in China for Windows 95? One copy."

37. Unfortunately, they also have a great deal of trouble understanding the idea that software should cost some something!

38. Steven Levy, *Crypto* (New York: Random House, 2000).

39. See Mark Stefik, "Shifting the Possible: How Trusted Systems and Digital Property Rights Challenge Us to Rethink Digital Publishing." *Berkeley High Technology Law Journal* 12 (1997): no. 1.

40. I recently searched Amazon.com for books about the English Levellers and Gerrard Winstanley. Only two of the twenty-three books about the Levellers are currently in print: none of the books about Winstanley are still in print. Without access to libraries, any research on these topics would end very quickly.

41. Stefik, "Shifting the Possible."

42. Stefik, "Shifting the Possible."

43. Tuck, *Natural Rights Theories*, 6.

44. Users program the VCR to tape a broadcast so that they can view the program at a more convenient time. Generally, the copy is immediately erased through reuse.

45. Ginsburg, "Putting Cars on the Information Superhighway," 1479.

46. Litman, "The Exclusive Right to Read."

47. Hesse, *Publishing and Cultural Politics*, 130-132.

48. NII White Paper, 73n.

49. NII White Paper, 84.

50. The *White Paper* authors assume that someone will argue that "fair use" has the same meaning as "use in education." But the concept of fair use involves the advancement of *knowledge*, a much narrower pursuit than the advancement of education *per se*. The *White Paper* authors have already spilled a great deal of ink rejecting any specific understanding of the concept of fair use, then claim that they will be exploited because the concept of fair use has no specific limits. See Samuelson, "Is Information Property," for a more detailed analysis on this point.

51. Yen, "Restoring the Natural Law," 551.

6

Commons: The Third Form
Of Property

Philosophers have been attracted to the idea of a commons for a number of reasons, both theoretical and metaphorical. A commons seems to presuppose a kind of egalitarianism: no commoner has an *a priori* claim greater than any other. Second, a commons seems to make sense even in the absence of political structures: accordingly, it has been seen as the sort of raw material out of which a political order can be formed. To the most radical, like Gerrard Winstanley (1609–1672), the commons is seem as a *solution* to the problems of distributive justice: men need to return to an initial state of innocence where "mine and yours" are no longer important.[1]

Much of the classical debate over the commons turns on whether or not community is the original state of humanity, followed by emergence of private property. If so, we can see the transition as either a providential concession to human wickedness (the position of Aquinas and Suarez) or the means by which humanity climbed beyond bare subsistence. The classical discussion was also complicated by actual ongoing transition between a feudal patronage-based economy and a modern money-based economy.[2]

We will begin by providing a rationale for appealing to a theory of the commons. We will then examine two sets of objections to the commons concept. The first set are "external": they challenge the coherence and relevance of the commons concept to understanding property rights. The second set is "internal": these problems are created by the logic of the commons itself. This set of problems helps us identify essential constraints that must be part of a commons regime. In the next chapter we will look at several historical examples that illustrate the problems of the commons and some potential solutions.

Prelude: The Vision of a Common World

Imagine a vast world, full of resources, in which there are only a few people. The people live together in small towns, surrounded by the vast, fertile world. The people need the resources of the land to survive, but there is always enough for more people. The people use the resources of the land to make things to use or sell. When someone cultivates a field it becomes "his" field while he is using it. But because there is no scarcity, the idea of claiming land to yourself seems absurd. Why lay claim to more than you can use? You have nothing to gain from such covetousness, and nothing to lose by foregoing an exclusive claim.

In this world, people use things while they need them, then simply give them up. Everything comes from the land, and everything eventually is given back to the land when no longer needed. Boundaries between "mine" and "yours" are set by our current use alone, not by the mere history of my use. Since everyone dies eventually, everything must eventually be given up.

Some people make very minimal use of resources. Others make very heavy use, but there is so much land and so few people that there is no depletion of resources. Since there is no scarcity, there is no pressure to increase productivity. Since no one gains more than subsistence from their labor, people only labor enough to achieve subsistence. A few, driven by their own curiosity or altruism or desire for fame, do innovate: some of these innovations become the basis for yet more innovations.

Some of the people find the idea that everything must eventually be given up disheartening: they see no reason to do more than what is necessary to meet their immediate needs. Others see this giving-up as a chance to give something back to the world and to their communities. They strive to leave everything better than they found it. Those who work have two rights: the right to the use of what they have made while they need it, and the right for their own unique contributions to be recognized.

Rationale for Introducing the Commons

Why should we appeal to the concept of a commons? Though the concept was popular in the past, the ideological collapse of Marxism has made suspect any analysis of ownership not derived from individual rights. The arena of debate seems to allow for two systems of ownership: individual ownership and state ownership. However, we should not be willing to simply define "common" as a proxy for "state." A commons theory does not need to be collectivistic in the sense of demanding that individual rights must be derived from community rights. I would go further and argue that a commons theory *can't* be collectivist, since the logic of collectivism is simply a mirror-image of the logic of individualism. The type of theory we need to ground IP rights must allow a commons to coexist harmoniously with private property.

I see the following advantages in exploring the idea of a commons :

- A coherent theory of the commons creates a "third way" between private and public ownership. There is at least the possibility of escaping the zero-sum, tug-of-war perspective created by dividing the world into two disjoint sets ("private" and "public").
- A theory of the commons allows us to explore less exclusive forms of property rights. Commons theories can handle overlapping and partial property rights in a more natural way than a theory that takes full liberal ownership as the foundation of property rights.
- The commons more naturally accommodates a pluralistic theory of property rights. It is not necessary to perform intellectual gymnastics to derive social rights from individual rights or *vice versa.*
- Finally, the commons suggests that it might be possible to talk about the origins (and therefore the legitimacy) of property rights in terms of some conceptually prior moral framework. An individualistic theory of property sees no connection between appropriation and morality until scarcity emerges. A commons-oriented approach can raise global moral issues from the beginning, with potentially non-individualistic consequences on issues of distributive justice.

Of course, it will do no good to build a commons-based theory if the idea of a commons is fatally flawed. We will need to answer a number of objections even to get started :

- Appealing to the commons is simply a distraction from dealing with the conflict of individual rights. For even if the resources of the world are "common," it is *individual* acts of labor and appropriation that actually sustain people.
- If there ever *was* an age of the commons, it is surely over now. Everything that is either belongs to individuals or to states.
- The idea of a commons simply reduces to *res nullius* (things owned by no one), and has no positive content. Appropriation of *res nullius* has no moral implications.
- Commons are transitory or unstable systems that must inevitably be replaced by private or collectivist ownership (this argument is known as "the tragedy of the commons"[3]). According to this argument, "common" just means "abundant." Commons make no sense in situations of scarcity, since they provide no principled criterion for favoring the interest of one commoner over another.

Assuming that we make it past those objections, we will need to deal with a set of issues internal to the commons approach :

- If we want the commons to coexist with private property, we need some way to prevent a commoner from simply claiming the whole commons at once. I call this *The Conquistador Problem.*
- If we want to understand the commons as in some sense the "joint property" of the commoners, we need to find some way that commoners can appropriate without having to get the consent of all the other commoners every time. Following Sreenivasan, I call this *The Problem of Consent.*[4]
- Common property regimes need systematic ways to address issues associated with scarcity, since scarcity means the *de facto* exclusion of some commoners from the resources of the commons. Scarcity introduces differences between "early arrivers" and "latecomers" that seem at least initially to be morally arbitrary. I call this *The Problem of Scarcity.*
- One possible outcome of scarcity is the end of the common-property regime by depletion or some systematic process of privatization. *The Problem of Change of Regime* deals with making such a transition in a fashion consistent with principles of distributive justice.

Finally, the concept of an intellectual commons involves dealing a further objection:

- The metaphor of an *intellectual* commons is inherently absurd. Common things are common precisely because they have an existence outside of human activity: people find the commons, not build it. But is there anything more consummately "artificial" than the Internet or the Library of Congress?

Chapter 7 will address the artificiality of the intellectual commons. Here we will argue that these immediate objections can be met, and move onto analyzing the internal problems of the commons concept. We will begin this chapter with some definitions, and then explore the conceptual objections and internal issues of a theory of the commons.

Basic Definitions

A *commons* consists of three sets:

- A domain of action that does not belong to any individual or group (in the sense of full liberal ownership). This is the commons itself.
- Objects that can be created or extracted from the resources of the commons. We will refer to these as the *fruits* of the commons. The commons is considered disjoint from the set of fruits.
- A set of agents that can make use of the commons. Following the seventeenth-century convention, we will refer to the members of this set as *commoners*. The set of commoners can be either open or closed, but either way we will assume that the commoners are not joint tenants (i.e., they do not collectively possess the rights of full liberal ownership). The point is vital enough to be repeated: *commoners' rights are use rights, not full liberal ownership* (see appendix B).

The restrictions on ownership and joint tenancy are necessary to prevent the commons from immediately collapsing into some form of private ownership. Note also that the definition is consistent with the coexistence of commons and a regime of private ownership. We are simply assuming that the existence of private ownership does not imply that everything is privately owned.

Given a commons, we can identify several distinction modes of use :

- **Innocent Use:** A commoner uses the commons in such a way that his use has negligible impact on the ability of other commoners to use the commons. An example would be walking across the commons to get from one place to another.[5]
- **Common Use:** A commoner consumes fruits of the commons while not substantially reducing the productive capacity of the commons. Examples include gathering fruit, hunting game, and planting and harvesting plants.
- **Destructive Use:** A commoner takes ownership of fruits of the commons while simultaneously reducing the productive capacity of the commons. An example would be picking fruit off a tree and then chopping it down, or depleting a non-renewable resource. Uses that are individually ordinary can become collectively destructive, for example overgrazing.
- **Enclosure:** A commoner asserts full liberal ownership over a part of the commons, thus diminishing the commons itself. We will refer to this as *enclosing appropriation*.

We make no claim that the list above is exhaustive: these are simply the modes of use that are most theoretically significant. It should also be noted that only enclosure is conceptually tied to the concept of a common. The other modes make perfect sense when applied to individual ownership or joint tenancy. A more significant problem with the definition above is its negative character: in effect, we are presupposing the concept of private property (if only to state that common rights are something different). We will be looking for more positive ways to describe common rights as the chapter progresses.

Let us also define a regime that, while somewhat similar to a commons (in terms of its beginning) evolves in a very different way. A *frontier* is a commons that is common by default, usually because it is newly discovered or previously unoccupied. A frontier differs from a commons in that it is *intended* to be enclosed into private holdings. The frontier "vanishes" when it has been totally enclosed.

At the opposite extreme is what we will call a *strict* common. In a strict common, only innocent use is permitted. An example would be a national park or a nature preserve. A strict common has been exempted from being an economic resource (one thinks of the often-seen slogan "take only photos—leave only footprints").

Appropriation, Enclosure, and Occupation

We can distinguish between two methods of creating property from a commons: appropriation and enclosure. Appropriation involves taking something from the commons and claiming it as one's own. The paradigm example from English history is gathering fuel from the commons (typically, sticks and cattle dung). The commoner removes a resource from the commons and consumes it. The other method we will call enclosure: a commoner "walls off" a fully functional piece of the commons itself for his exclusive use. The distinction we are after here is a distinction between diminishing the resources of the commons and diminishing the commons itself.

Though our theoretical work would be significantly more straightforward if these categories were disjoint, we must accept that they are not. The commons can be rendered useless by common overuse just as surely as it can be destroyed by enclosure. We will recognize this possibility by dividing use into three forms: non-consumptive use, occupation, and consumptive use. Non-consumptive use is use that does not significantly compromise the availability of resources for commoners. It leaves, in Locke's famous phrase "as much and as good" for the other commoners.[6] The process of appropriation itself requires that appropriation creates an exclusive right to resources *after their appropriation*. It does me no good to gather firewood if every other commoner has the right to "gather" firewood by simply taking it from me. The classic example is one given by Cicero and Seneca: a seat in a theater belongs to no one while it is empty, but becomes "my" seat while I sit in it.[7] We will refer to the exclusive right during use as *occupation*.

Consumptive use is use that significantly compromises the availability of resources for other commoners. Some forms of appropriation are inherently consumptive: mining and oil extraction inevitably reduce the resources available to other (or future) commoners. Non-consumptive use can also become consumptive: paradigm examples would be overgrazing, deforestation, or pollution.

Furthermore, the distinction rests on the ability to distinguish between the resources produced by the commons and the commons itself. This is not always straightforward, especially in cases like overgrazing, where appropriation by commoners destroys the productive capacity of the commons. The moral I see here is something like this: the distinction between appropriation and enclosure is not practical, but moral. Though the boundaries may be fuzzy, there is a clear distinction between creating property through use ("spending income") and creating property by diminishing the commons itself ("spending capital").

The distinction is especially problematic for IP because the "space" of the intellectual commons is not finite, and *all* boundaries seem to be purely conventional.

We need to further clarify what happens when a commoner uses the resources of the commons. When I pick an apple from a common apple tree, does the apple become my private property? Suppose I pick a pile of apples. Would another commoner have the right to "pick" apples from my pile rather than the tree? If not, why not?

Both Grotius and Pufendorf distinguished between common use (which they called "occupation" and private ownership. They found the distinction important because they wanted to argue that common use preceded the institution of private property. Grotius puts it as follows:

> Soon after the creation of the world, and a second time after the flood, God conferred upon the human race a general right over things of a lower nature. "All things", as Justin says, "were the common and undivided possession of all men, as if all possessed a common inheritance." In consequence, each man could at once take whatever he wished for his own needs, and could consume whatever was capable of being consumed. The enjoyment of this universal right then served the purpose of private ownership; for whatever each had thus taken for his own needs another could not take from him except by an unjust act. This can be understood from the comparison used by Cicero in his third book "On Ends": "Although the theatre is a public place, yet it is correct to say that the seat a man has taken belongs to him."[8]

Pufendorf argues that common use is not ownership, since according to his theory property does not exist prior to the emergence of society, but he points out that "such universal use of things somehow served in place of ownership."[9]

Natural Law and the Changing Nature of Property

The rhetorical use of the commons by natural law theorists followed a systematic pattern. A state of nature defined in the kind of schematic manner given above was postulated. Then the theorist would identify various moral problems, and adduce moral principles that would both fill in the ethical gaps and also arise "naturally" from the logic of the commoner's situation.

Natural law theorists wanted to show, among other things, how property could develop out of the state of nature. This was of much greater theoretical than historical interest: a plausible process would show how distributive justice could be built "from scratch," or at any rate from the nature of people and the logic of their original situation.

As one might expect, such a theory sees the commons differently at different times. Grotius, Pufendorf, and Locke all describe a gradual abandonment of the commons in favor of private property. This put them at odds with an absolutist like Filmer, who saw an unbroken line of *dominium* stretching from Adam to their own time.

The change that the natural law theorists pointed out has two engines. The first, more obvious one is the proliferation of claim rights generated by the existence of more than one commoner. "Man may be endowed communally, but he must be nourished individually. Yet in taking any particular item from the common, it would seem that a man violates the rights of other commoners, to whom *ex hypothesi* that particular item also belongs."[10] The problem existed for Adam and Eve, but becomes more and more acute as the children of men proliferate.

The second engine is a constraint imposed by the physical world: finitude, and its inevitable consequence, scarcity. Scarcity ultimately breaks down the commons itself, as everything becomes occupied or consumed. Eventually every seat in the theater is occupied, and the next commoner finds that the commons itself has disappeared.

We will begin by exploring the problems identified above.

Only Individuals Matter

Someone like Robert Nozick would reject the idea of starting anywhere *but* with individual rights. He states this position at the beginning of *Anarchy, State and Utopia*:

> Individuals have rights, and there are things no person or group may do to them without violating their rights. So strong and far-reaching are these rights that they raise the question of what, if anything, the state and its officials may do. How much room do individual rights leave for the state? . . . Our main conclusions about the state are that a minimal state . . . is justified; that any more extensive state will violate person's rights not to be forced to do certain things, and is unjustified.[11]

This minimalism leaves no room for a positive conception of the commons (or for any other entity that would constrain individual rights). For Nozick, the commons is simply the set of unowned things: its only purpose is to provide a foundation for original appropriation. Even his discussion of the "historical shadow" of the Lockean proviso is motivated by a desire to avoid the "moral catastrophe" of violating *individual* rights wholesale: "the Lockean proviso is not an 'end-state principle': it focuses on a particular way that appropriative actions affect others, and not on the structure of the situation that results."[12]

Yet even Nozick realizes that a "purely private" world would be problematic. He makes the following observation:

> The possibility of surrounding an individual presents a difficulty for a libertarian theory that contemplates private ownership of all roads and streets with no public ways of access. A person might trap another by purchasing the land around him, leaving no way to leave without trespass . . . Whatever provisions he has made, anyone can be surrounded by enemies who cast their nets widely enough.[13]

How can we prevent such a situation from arising? Not by leaving unappropriated land for the roads, since they can simply be appropriated by the next person who comes along. Nozick asserts that the Lockean proviso would prevent this situation from arising, but how? The status of the mandated "right of way" *cannot* be merely *res nullius.* Nor can it be any limited partnership, since this would still leave the partners free to deny the rights of non-partners. The only approach that absolutely precludes the type of injustice Nozick is worried about involves treating the right of way as a public good. Even a libertarian world can't simply divide everything into the sets "owned" and "not yet owned."

Commons Only Matter in the Beginning

Though the property theories developed by Grotius, Pufendorf, and Locke are different in various ways, they all share a kind of "etiological" flavor. All three tell plausible stories that explain how men started in a world held in common and ended up in a world dominated by private property.[14] Only Grotius seems to have an ongoing role for a commons in his account of property (and that in unconfinable entities such as the sea). Pufendorf believes that the commons ends when the state of nature ends, and Locke argues that the disappearance of the commons is ultimately irrelevant, since almost all value is created by labor. Locke's attitude especially suggests that the commons matters only at the beginning of human society.[15]

This conclusion cannot be correct for the reasons suggested above. Any coherent theory of property seems to require the continuing existence of a domain of things available for use but not available for private ownership.

We will evaluate the merits of this argument when we discuss the problem of change of regime. Here we are simply making the point that it is certainly not *obvious* that allowing the commons to disappear makes no difference to the commoners. A strong conclusion needs the support of a strong argument. We do well to scrutinize candidates carefully without assuming that change of regime can be effected in some automatically just way.

The Idea of a Commons Has No Positive Content

We will consider both a naive and a sophisticated argument that the commons is a mere negation. It could be objected that commons-talk either refers to things beyond ownership, or things that, as a matter of fact are not yet owned. In the first case, commons-talk is irrelevant to any real issues about property: in the second case, it is simply a contingent historical fact that may (or may not) have some relevance to justice.

In addition to begging the question, this objection confuses theories of property with rules for alienation of property. Our real estate laws don't have to cover the conditions of ownership of intrinsically unownable things, but in many cases it is the *identification* of a particular thing as unownable that is the issue at hand. Before we address issues of distributive justice it may be appropriate to ask whether the thing in question should be owned at all. In either case, we need to consider the possibility that some kind of non-individual ownership might make sense even when individual ownership does not.

We will now consider a more subtle line of thought. Samuel Pufendorf famously argued that the commons can only be understood negatively:

> a right to all things, antecedent to any human deed, is not to be understood *exclusively*, but *indefinitely* only; that is, we must not imagine one may engross all to himself, and *exclude* the rest of mankind; but only that nature has not *defined*, or determined, what portion of things shall belong to one, what to another, till they shall agree to divide her stores amongst them, by such allotments and divisions.[16]

> things are said to be negatively common, as considered before any human Act or Agreement had declared them to belong to one rather than another. In the same sense, things thus considered are said to be *No Body's*, rather negatively than privatively, i.e. that they are not yet assigned to any particular person, not that they are incapable of being so assigned. They are likewise termed Things that lie free for any taker.[17]

Tully argues that Pufendorf's identification of the commons with the currently unowned ultimately reduces all property to private property: "To say that property cannot belong in the same manner and in whole to more than one person is to deny that common ownership is a form of property. . . . The notion that property is, *ipso facto*, private property passes from here into eighteenth-century Europe through the widespread use and republication of the writings of Grotius and Pufendorf."[18] Pufendorf's view also implies that it is not possible to say anything particularly interesting about the commons *sui generis*, since it is only defined in reference to private property.

Again, our response is to argue that the existence of public goods and inherently non-exclusive resources shows that Pufendorf's analysis cannot be the whole story of the commons. In effect, Pufendorf is "solving" the problem of common ownership by legislating it out of existence. In the absence of any theory that can explain public goods in terms of negative community, we must reject Pufendorf's argument.

Commons Are Inherently Unstable and Transitory (the Tragedy of the Commons)

Many philosophers are familiar with Garrett Hardin's paper "The Tragedy of the Commons."[19] Hardin argues that systems of what are now called "open-access resources"[20] are inherently unstable. The problem is that each commoner experiences the full benefit of using up more resources but only experiences part of the costs of that consumption. If we assume that commoners are motivated only by rational self-interest, the expected outcome is rapid depletion of all available resources. Hardin and his followers have seen only two ways out of this tragedy: private property or the development of an authority that uses coercive methods to enforce global consumption limits ("Leviathan"). Private property prevents the tragedy by forcing each (former) commoner to bear the full cost of his own overconsumption: presumably this would lead them into taking better care of what they have. Either way, a common property regime must ultimately collapse or be abolished.

Recently economists have begun exploring the mirror-image of the tragedy of the commons: the tragedy of the anticommons.[21] Imagine a world totally divided into individual holdings. In order to get my goods to market, it might be necessary for me to cross the property of dozens or even hundreds of owners. If all of them want a piece of the action, the cost of commerce could become so great that it would break down entirely. Even more dire consequences are possible: where would the poor even stand? Anticommons can also arise in situations where property rights overlap on a large scale: reconstruction of the city of Kobe after a massive earthquake has been hampered by the fact that land in the city is subdivided into plots as small as two square meters, and many of these plots are subject to several different ownership rights.

Both tragedies seemed to have been largely unnoticed by classical theorists of property (though Aristotle did point out that common things are cared for less than individual things). Both Locke and Rousseau could gloss over them by their appeal to the vastness of the world: with an empty enough world, we need not expect inevitable conflict. We, however, do not have the same freedom: there are few (if any) unclaimed places left.

I would argue that Hardin has not proved the *impossibility* of a commons, but rather its *necessity*. The common message of these tragedies is also the message of the prisoner's dilemma: even totally self-interested agents cannot maximize their self-interest by ignoring the preferences and behavior of others. An agent's self-interest can only be served by occasionally looking beyond self-interest. Individual property and the pursuit of wealth rest on a scaffolding of resources that are not individual property. We cannot afford to treat this common dimension as a mere negation: we must understand its positive reality.

Problems of the Commons

The basic model of a commons described above is incomplete in a number of ways. In this section we will examine four potential problems with the commons concept. These problems point to various ways in which political rules need to be incorporated into the commons model to make it more viable.

The Conquistador Problem

The first problem concerns a certain tension between the commons concept and the process of appropriation. Suppose that I discover a new island and decide to emigrate there. Since the island is uninhabited and previously undiscovered, it belongs to no one. I could decide to claim some land, then clear it and cultivate it. But if no one owns any of the island at all, why can't I simply claim ownership of the whole thing? The same problem would seem to apply to a more orthodox kind of commons. If a commoner has the right to appropriate apples, why doesn't he have the right to appropriate *all* the apples, as well as the commons itself? If this argument has any merit, then it would seem that any commons that tolerates unlimited appropriation on the part of commoners would collapse as soon as the first aggressive claimant (a "conquistador" arrives).

Rousseau makes a similar argument in *The Social Contract*, in the course of arguing against a claim-based theory of appropriation:

> Is it possible to leave such a right unlimited? Is it to be enough to set foot on a plot of common ground, in order to be able to call yourself at once the master of it? Is it to be enough that a man has the strength to expel others for a moment, in order to establish his right to prevent them from ever returning? How can a man or a people seize an immense territory an keep it from the rest of the world except by a punishable usurpation, since all others are being robbed by such an act, of the place of habitation and the means of subsistence which nature gave them in common? When Nunez Balboa, standing on the sea shore,

took possession of the South Seas and the whole of South America in the name of the crown of Castile, was that enough to dispossess all their actual inhabitants, and to shut out from them all the princes of the world? On such a showing, these ceremonies are idly multiplied, and the Catholic King need only take possession all at once, from his apartment of the whole universe, merely by making a subsequent reservation about what was already in possession of other princes.[22]

Rousseau's point is that just appropriation requires more than force and also more than the mere act of claiming ownership. While the scope of his argument is broader than the one we are considering here (since he considers expropriation as well as appropriation) it applies equally well to the situation of commoners.

This problem was noted even earlier by Pufendorf in his discussion of occupation: "For if one man, for example, had been conveyed with his spouse to a vacant island sufficient for supporting myriads of people, it would be impudent for him to claim the whole thing for himself by virtue of his title of occupancy, and try to expel those who landed on a different part of the island."[23]

The problem is presented in its starkest form by Hobbes. In *Leviathan*, Hobbes argues that in a state of nature *each* commoner has a right to *everything* in the commons.

> *right* consisteth in liberty to do, or forebear; . . . And because the condition of man (as hath been declared in the precedent chapter) is a condition of war of every one against every one, in which case every one is governed by his own reason, and there is nothing he can of that may not be a help to him in preserving his life against his enemies; it followeth that in such a condition every man has a right to every thing, even to another's body.[24]

Or consider a more modern manifestation of the same problem. The free software foundation releases its software under the "Gnu Public License," which states various conditions of use and copying. The fact that they impose a multipage contract on software which they would like to be "free as air"[25] might seem puzzling until you consider the consequences of releasing the software without any conditions at all. If the Free Software Foundation simply published their programs without conditions, there would be no reason why someone else couldn't make some trivial modification of the code and then copyright it himself. The new "author" would then be able to sue the Free Software Foundation for infringing "his" copyright!

The moral of all these examples is that purely unrestricted appropriation is not consistent with a commons regime. In order for all commoners to have the right to the use of the commons, some prior limitations must be imposed on appropriation. Rousseau offers three:

- the item to be appropriated must not already be owned by someone

else
- "a man must occupy only the amount he needs for his subsistence"
- "possession must be taken, not by an empty ceremony, but by labour and cultivation, the only sign of proprietorship that should be respected by others"[26]

Rousseau's solution is similar to Grotius' appeal to the concept of occupation. In both cases, natural limits on the ability to labor and consume naturally limit appropriation.

It is also possible to use a teleological principle to solve the conquistador problem. Locke addresses the problem by imposing a restriction on appropriation that Sreenivasan calls the spoilage limitation.[27] Locke states it as follows:

> It will perhaps be objected to this, That if gathering the Acorns, or other fruits of the Earth &tc makes right to them, then any one may *ingross* as much as he will. To which I answer, Not So. The same Law of Nature, that does by this means gives us property does also *bound* that *Property* too . . . But how far has he [God] given it to us? *To enjoy.* As much as any one can make use of to any advantage of life before it spoils; so much he may by his labour fix a Property in. Whatever is beyond this, is more than his share, and belongs to others. Nothing was made by God for Man to spoil or destroy.[28]

The spoilage limitation solves the conquistador problem by "scaling down" the commoner's appropriation to match his own needs and his own capacity to labor. The limitation imposed is both moral and practical. A commoner is only morally justified in appropriating what he can use without wastage. A commoner is *practically* constrained by his lack of control over the other commoners to appropriate only what he can immediately use. Any other claim is merely "empty ceremony."

The spoilage limitation also explicitly denies Hobbes' conclusion that every commoner has a right to everything in the commons. Even if a commoner's power should exceed his needs, he lacks the ability to use the entire commons himself without wastage.

We will consider possible justifications of the spoilage limitation in the next chapter. Here we will simply mention two possible lines of argument. It would be possible to argue, as Locke does above, along theological lines: God's purpose in giving the commons to man is man's preservation. Wastage defeats God's generosity and thus cannot be justified. A second possibility would be to justify the spoilage limitation in terms of the commoner's rights: it could be argued that when a commoner wastes the resources of the commons he is doing an injury to the other commoners. As we shall see in the next chapter, such an argument can raise difficult issues about overriding existing property rights.

The Problem of Consent

One hasty response to the Hobbesian form of the conquistador problem would be to claim that the commons and its fruits belongs *jointly* to the commoners. If the commons is some form of joint ownership, then it is not the case that every commoner has a right to the entire common, which rules out the possibility of conquistadors.

Unfortunately, this approach replaces one problem with another. It would appear that before one commoner can consume part of the commons, he is obligated to get the consent of all of the other commoners. If this is the case, then the commons ceases to be any kind of arena for the free action of agents, and in an important sense is no longer "common."

Grotius was aware of the problem of consent, and tried to deal with it by arguing that each commoner could count on the *tacit* consent of other commoners. When describing the emergence of property from community he writes; "This happened not by a mere act of will, for one could not know what things another wished to have, in order to abstain from them—and besides, several might desire the same thing—but rather as a kind of agreement, either expressed, as by division, or implied, as by occupation."[29]

In *The Law of Nature and Nations*, Pufendorf argues that dominion analytically presupposes agreements between people: "We must, before dealing with it, preface that ownership and communion are moral qualities that do not physically and intrinsically affect things themselves, but produce only a moral effect in relation to other men, and that these qualities—like the rest of this sort—refer their birth back to imposition."[30] While there was only one person, the concept of ownership was meaningless: "Furthermore, since positive community and ownership involve a relation to other men, it is not at all accurate to say all created things were Adam's own, but only this, that he was the owner of all things in a concessive but not in a formal sense, that is insofar as there was no one's right to hinder him from being able to convert all things to his own use if there was a need."[31]

For Pufendorf, the community of God's original grant to humanity is purely negative:[32]

> Things are said to be common in the former manner [negative] insofar as they are considered previous to any human deed which declares them to belong more especially to this person than to that. They are also, in this sense, said to be *no one's*, that is, in the negative sense of not yet having been assigned to anyone in particular rather than in the privative sense of being incapable of such assignation. And they are referred to as common stock available to all. Things common in the other sense [positive community] however, differ from those that are one's own in this point only, that they belong to several persons in the same manner while the latter belong only to one.

While both Grotius and Locke are committed to positive community as the primordial state of man, Pufendorf argues that God has left the nature of community entirely in the hands of humanity. "God did, indeed, allow man to take the earth, its products and animals for his own use and advantage, giving him an indefinite right to these things. Yet, the manner, degree and extent of this authority were left to man's discretion and disposition."[33]

Locke took another approach: he argued that there are conditions under which appropriation without consent is morally permissible. As long as a commoner can leave "as much and as good" for the other commoners, his appropriation cannot be said to harm the other commoners. However, this raises yet another problem: what about cases where leaving as much and as good isn't possible?

The Problem of Scarcity

Part of the appeal of a commons is that it's—well—*common.* There is no difference between the commoners in terms of their right to use of the commons. Those concerned with issues of maximum freedom and equality could not do better.

The problem is that there is a kind of entropy in a commons: it becomes less and less valuable over time. The decline can be due to destructive use, such as overgrazing or deforestation: or it can be due to the fact that a finite commons inevitably inevitably has limited resources. A theater only has a finite number of seats. The pasture only has a finite amount of grass. When all the seats are full, it really doesn't matter that the next person has a "right" to a seat, because there aren't any left.

Our first question is whether or not positive community can survive the emergence of scarcity. Grotius argued that it could, and Locke agreed with him.[34] Grotius argued that in cases of dire need, a person has the right to consume or destroy the surplus property of another. The "benign reservation" is that it was the intention of those who first accepted private ownership to "depart as little as possible from natural equity."[35]

Grotius is typically cautious in hedging this right of necessity with conditions: the necessity must be unavoidable,[36] the right does not apply against someone in equally dire need,[37] and justice requires restitution after the emergency, if this is possible.[38] The right of necessity applies in situations of general scarcity as well: "Hence it follows, again, that on a voyage, if provisions fail, whatever each person has ought to be contributed to the common stock."[39]

As we have seen, the internal coherence of the commons requires that all commoners have equal rights over the commons. This is straightforward when there is some bound on appropriation (to avoid the conquistador problem), and when available resources exceed the demands made by commoners. The situation becomes much more complicated when demands by commoners exceeds resources. In a situation of scarcity, it is not possible for all commoners to have equal access. If the show is sold out, then I can't have a seat. Yet what justifies the difference between those in the theater and those outside?

The problem is that the commons model seems to favor those who get there first. Yet there is no difference between the early bird and the late-comer *except* their time of arrival.

Are All Finite Commons Frontiers?

One solution would be to declare the commons regime ended when all available resources are in use. This approach sees the commons as a frontier: when everything has been claimed, the commons is gone. It also follows that commons can only begin and be sustained in situations of abundance: when scarcity sets in, a new set of rules starts to apply. This interpretation makes sense if we interpret the commons as negative community: "belongs to no one (yet)" as opposed to "belongs to everyone."

There are two problems with this approach. First, it makes the commons either a sort of moral luxury permitted by the purely contingent facts of supply and demand, or merely a situation in which we can afford to ignore the "real" (i.e., private) distributive issues because there is so much to go around. On either interpretation, the commons is treated as an initial condition rather than a moral regime.

Second, it seems to assume that rights only move in one direction, from the commons to private property. This assumption seems plausible as long as we only consider consumptive use, but it does not fit other possible regimes. It could be that individually appropriated resources could revert to the commons, or that the commons itself could be enlarged.

The worst problem created by scarcity is the *de facto* creation of two classes of commoners: those who got to the commons first (and thus were able to appropriate), and those who arrived too late. Why should the first group have rights the second group can't have?

The Problem of Latecomers and Change of Regime

Nozick raises the problem of latecomers as follows: "Is the situation of persons who are unable to appropriate (there being no more accessible and useful unowned objects) worsened by a system allowing appropriation and permanent property?"[40] He sees this question as equivalent to questioning the legitimacy of private property, and attempts to answer it by appealing to the positive utility of private property, but concedes that a really complete answer would require some way of quantifying the value of access to the commons.

According to Karl Olivecrona, Locke's labor theory of value was meant "to belittle the grounds for such complaints."[41] Locke can respond to the problem of latecomers by pointing out that since virtually all of the value of things is created by labor, "Man (by being Master of himself and *proprietor of his own person and the actions or labor* of it) had still in himself *the great foundations of* Property."[42] It is unclear that Locke was justified in being so unconcerned about the disappearance of the commons. Gopal Sreenivasan argues that neither the Lockean proviso nor the labor theory of value are enough to preserve the original equality of rights:

> Landless commoners are not at liberty even to produce a surplus, since their access to the necessary materials depends on the permission of the landowners. Furthermore, where they are permitted to produce a surplus, not only do landless commoners have no claim-right to keep most, or perhaps even any, of the surplus they produce, but they are not at liberty to keep it either. The benefit of labour's abundance—which Locke so celebrates—is therefore placed at the exclusive disposal of the landowners. . . . Clearly, then, the natural property regime fails to preserve the access to the materials necessary to produce the means of comfort and support to which each commoner was originally entitled.[43]

The only really straightforward solution to the problem of latecomers is a drastic one: deny that appropriation creates full liberal ownership. If appropriation creates only use rights, then no commoner's access is threatened by appropriation. A (slightly) less drastic solution would be to interpret every commoner as having a right to access the commons "through" another commoner's property. In either case, a common property system *cannot* create exclusive rights to the commons itself.

In what follows, we will assume that some form of common property regime may be useful in understanding intellectual property. This commons is not simply the complement of the set of all owned things. It includes things that cannot be owned, and things whose individual ownership is inconsistent with social life.

The commons concept not only has content: its content is pushed in a consistent direction by our analysis of the problems outlined above. These problems force a commons theorist to either limit or deny the possibility of ownership of the commons itself (though not necessarily the fruits of the commons). Common rights are naturally *use* rights rather than exclusive ownership rights. Even when a commons theory allows appropriation, that appropriation must be bounded by some natural constraint (to avoid the conquistador problem).

Notes

1. Christopher Hill, *The World Turned Upside Down: Radical Ideas During the English Revolution* (London: Penguin Books, 1972).
2. One of the demands of the Levellers during the English revolution was that all land titles and land transfers be legally recorded, thus forcing patronage and prerogative into the open. See Andrew Sharp, ed., *The English Levellers* (Cambridge: Cambridge University Press, 1998), 95 (Clause 4 of the Agreement of the People).
3. Garrett Hardin, "The Tragedy of the Commons," *Science* 162 (1968): 1243–1248.
4. Gopal Sreenivasan, *The Limits of Lockean Rights in Property* (Oxford: Oxford

University Press, 1995).

5. Hugo Grotius, *On The Laws of War and Peace (De Jure Belli ac Pacis*, hereafter DJBP) (Oxford: Carnegie Endowment, 1923), 2.2.1.1. We will use the seventeenth-century "dotted number" method of citation, which is edition independent. "2.2.11" accordingly means book 2, section 2, chapter 1, paragraph 1. Samuel Pufendorf uses three levels (section/chapter/paragraph), while Locke's *Treatise* uses two (treatise/paragraph).

6. Locke, *Two Treatises*, II.34.

7. Grotius, DJBP, 2.2.3.

8. Grotius, DJBP, 2.2.2.1.

9. Pufendorf, *The Law of Nature and Nations*, 4.4.9. Except where otherwise noted, all quotations from Pufendorf are from *The Political Writings of Samuel Pufendorf* (Oxford: Oxford University Press, 1995).

10. Sreenivasan, *Limits of Lockean Rights*, 25.

11. Nozick, *Anarchy, State and Utopia*, ix.

12. Nozick, *Anarchy, State, and Utopia*, 181.

13. Nozick, *Anarchy, State, and Utopia*, 55n.

14. Buckle, *Natural Law and Property*, 6-15.

15. Karl Olivecrona, "Locke's Theory of Appropriation," *Philosophical Quarterly* 24, no. 96 (1974): 220–234.

16. Pufendorf, *The Law of Nature and Nations*, 3.5.3. Quoted in Tully, *A Discourse on Property*, 74.

17. Pufendorf, *The Law of Nature and Nations*, 4.4.2. Quoted in Tully, *A Discourse on Property*, 74.

18. Tully, *A Discourse on Property*, 72-73.

19. Hardin, "The Tragedy of the Commons," 1248.

20. See Carol A. Rose, "Expanding the Choices for the Global Commons," for an explanation of the difference between open-access and common-property systems.

21. Francesco Parisi and Norbert Schulz, "Duality in Property: Commons and Anticommons." working paper 00-16, (2000), University of Virginia School of Law.

22. Rousseau, *The Social Contract*, 394.

23. Pufendorf, *The Law of Nature and Nations*, 4.6.3.

24. Hobbes, *Leviathan*, 86.

25. Richard Stallman, "Why Software Should Be Free," in Deborah Johnson and Helen Nissenbaum, eds. *Computers, Ethics, and Social Values* (Upper Saddle River NJ: Prentice-Hall, 1995) 190-199.

26. Rousseau, *The Social Contract*, 394.

27. Sreenivasan, *The Limits of Lockean Rights*, 34.

28. Locke, *Two Treatises*, II. 31.

29. Grotius, DJBP, 2.2.5.

30. Pufendorf, *The Law of Nature and Nations*, 4.4.1. Quoted in Tully, *A Discourse on Property*, 74.

31. Pufendorf, *The Law of Nature and Nations*, 4.4.11. Quoted in Tully, *A Discourse on Property*, 74.

32. Pufendorf, *The Law of Nature and Nations*, 3.5.3.

33. Pufendorf, *The Law of Nature and Nations*, 4.4.4.

34. Grotius, DJBP, 2.2.6.

35. Grotius, DJBP, 2.2.6.1.

36. Grotius, DJBP, 2.2.7.

37. Grotius, DJBP, 2.2.8.

38. Grotius, DJBP, 2.2.9.
39. Grotius, DJBP, 2.2.9.
40. Nozick, *Anarchy, State and Utopia*, 177.
41. Olivecrona, "Locke's Theory of Appropriation," 232.
42. Locke, *Two Treatises*, II.44.
43. Sreenivasan, *The Limits of Lockean Rights*, 113-114.

7

The Intellectual Commons

I sent you to reap a harvest you have not labored for.
Others have labored for it;
and you have come into the rewards of their labor.
                                    John 4:38

We will begin at the end: a rough sketch of my "geography" of the intellectual commons. I see the intellectual commons as naturally divided into three parts. The first is a protected zone that contains works that are still under IP protection. These works belong in the intellectual commons because some forms of use are acceptable, even during the time of protection. The second zone is what could be called the narrow intellectual commons. It contains works whose IP protection has expired, and are thus subject to greater use. The third zone is something completely different: it represents what could be called the greater intellectual commons, or the common knowledge of humanity.

The greater intellectual commons contains something other than works. It contains the "ideas" that are not legally protected, even though their "expressions" are. It also contains science, mathematics, and all other theoretical knowledge. It contains all common-sense knowledge and public historical facts (more on this distinction later). To the extent that archetypes exist, they would belong in the greater intellectual commons. All stock plots, clichés, and stereotypes are part of the greater intellectual commons.

Our legal norms explain how works move from the protected zone to the narrow intellectual commons. How do things get into the greater intellectual commons? Perhaps the term "getting into" is unfortunate: it suggests that the greater intellectual commons is another repository of works. The presence of something in the greater intellectual commons is a matter of its intrinsic nature rather than its history. The idea-expression distinction divides aspects of a single work into common aspects (the ideas) and aspects that can be owned (the expression).

Why can't the ideas in the greater intellectual commons be owned? There are a number of possibilities:

- They are ideas of such profound value to humanity that it would be unjust to allow anyone to assert property claims against them (Justin

Hughes calls these "extraordinary ideas." [1]
- They are ideas of such a pedestrian and obvious nature that it would be meaningless to assert property rights about them (Hughes calls these "everyday ideas").
- They are concepts which have a certain intellectual inevitability about them: they will be rediscovered time and again by people who need them.
- They are ideas that have been around for so long that their connection to a specific author and historical context has been lost in "time out of mind."
- They are ideas of such a high level of abstraction that they have no clearly delineated boundaries; the target of Learned Hand's "abstraction test."[2]

Must a Commons Be Natural?

The first and greatest conceptual hurdle any theory of the intellectual commons must overcome is the fact that it isn't "natural." Natural law theories have appealed to the idea of a commons as a foundation for property precisely *because* the commons is assumed to lie outside the scope of human projects. It is the unowned source of owned things, and the "pre-human" source of resources available to the independent person. How can we treat the products of human labor as a commons without ignoring the natural rights of their creators?

This objection makes two questionable assumptions: that everything produced by people is intrinsically private property, and that everything produced by people is owned. Both assumptions are false: at least some of the things produced by people do not seem to be meaningful objects of ownership (as when someone "makes history": they really make something, but not something they can sell), and many things produced by people are no longer owned by anyone (such as the contents of a dumpster).

Resolving the problem also requires the *assumption* that labor does not have to create a right to absolute ownership. We have already addressed this issue, but for the sake of clarity we should spell out some basic arguments. The question depends on the nature of the work and its relation to works created by other agents. Some things can be shared without loss, as the famous example of lighting a taper shows. Other works of the mind seem to have a desire to be free: they tend to "seep into" discourse and practice. Many works are actually composites of earlier works where we must avoid attributions of ownership on pain of a potentially infinite regress. Copyright theory is based on the assumption that the property status of works is never absolute (every work has both "owned" and "unowned" features) and that the property status of works can change over time. A labor theory of value needs to presuppose raw material unencumbered by previous labor, and thus must assume a commons that is not only unowned, but "natural" in the sense of existing independently of human effort. Movement is always *out* of the commons. But as we have noted, IP institutionalizes a movement of property *into* the commons through limitation of term. Thus IP theory is *ab initio* incompatible with a theory of property that assumes that labor always creates a claim right to property. This may seem odd, but it is not obviously absurd, and as we shall see it will help solve some problems that a simple labor theory of property cannot.

We will approach this problem slowly and cautiously, as one might approach a porcupine. One place to start is by looking at Arendt's concept of "public."

Arendt's Concept of the Public

In *The Human Condition*, Hannah Arendt develops what could be called a "practical-existential" approach to the commons.[3] Like Heidegger and Sartre, Arendt saw human reality as essentially constituted by the mutually disclosing presence of self and others: "Compared to the reality that comes from being seen or heard, even the greatest forces of intimate life—the passions of the heart, the thoughts of the mind, the delights of the senses—lead an uncertain and shadowy kind of existence unless and until they are transformed, deprivatized and deindividualized, as it were, into a shape to fit them for public appearance." The public realm is here taken to be both other people and the arena in which everything "can be seen and heard by everybody."[4] In this sense "public" is the necessary complement to "private" as an essential part of subjective existence.

Arendt identifies a second sense of the term "public" that is much more difficult to characterize, but seems to contain many of the features associated with the idea of a commons. As she puts it,

> Second, the term "public" signifies the world itself, in so far as it is common to all of us and distinguished from our privately owned place in it. This world, however, is not identical with the earth or with nature, as the limited space for the movement of men and the general condition of organic life. It is related, rather, to the human artifact, the fabrication of human hands, as well as to affairs which go on among those who inhabit the man-made human world together. To live together in a world means essentially that a world of things is between those who have it in common, as a table is located between those who sit around it; the world, like every in-between, relates and separates men at the same time. The public realm, as the common world, gathers us together and yet prevents our falling over each other, so to speak. What makes mass society so difficult to bear is not the number of people involved, or at least not primarily, but the fact that the world between them has lost its power to gather them together, to relate and to separate them.[5]

This second public is a domain of action as much as it is a domain of "appearance." It is also something more. It provides the only setting in which it is possible to transcend human mortality through a process of work and sharing:

> [T]he existence of a public realm and the world's subsequent transformation into a community of things which gathers men together, and relates them to each other, depends entirely on permanence. If the world is to contain a public space, it cannot be erected for one generation and planned for the living only; it must transcend the lifespan of mortal men. Without this transcendence into a potentially earthly immortality, no politics, strictly speaking, no common world and no public realm is possible. For unlike the common good as Christianity understood it—the salvation of one's soul as a concern common to all—the common world is what we enter when we are born and what we leave behind when we die. It transcends our lifespan into past and future alike; it was there before we came and will outlast our brief sojourn in it. It is what we have in common not only with those who live with us, but also those who were here before and with those who will come after us. But such a common world can survive the coming and the going of the generations only to the extent to which it appears in public. It is the publicity of the public world which can absorb and make shine through the centuries whatever men may want to save from the natural ruin of time.[6]

The public world is constituted by the actions and works of individual people, but it both predates and outlasts any individual life. It is not a commodity because no one can choose to opt out of it. Its existence and importance are inconsistent with the theory of valuational solipsism discussed in chapter 9.

The public world transcends mortality because some of the things people make acquire a "life of their own." Arendt makes a distinction between "labor" (activity to meet basic and repeating human needs) and "work" (the process of making something that acquires its own independent existence). These works can be physical objects or ideas that become part of the universe of discourse about the world. Though physical objects often outlast individual human lives, they are not immune to the natural ruin of time. Only ideas have that kind of permanence: stories, metaphors, and practical skills are sustained and elaborated, and as Arendt points out, only ideas that are "released" into the public world can have this kind of permanence. At the same time, this release of an idea ironically diminishes the significance of the author herself: as Foucault puts it, one result of the progress of science is that "The author function faded away, and the inventor's name served only to christen a theory, proposition, particular effect, property, body, group of elements or pathological syndrome."[7] Shakespeare is another famous example: though we possess a vast body of works that reflect an extraordinary mind and heart, virtually all the facts about the *person* William Shakespeare have disappeared. All that is left is his name, and his works, which have become part of the intellectual commons.

The Case for the Intellectual Commons

Fortified by the preceding meditations, we will now attempt to make a case for the intellectual commons.

The Raw Materials of Works by Nature Cannot be Owned

There are a number of good arguments for the thesis that ideas cannot be owned:

- Ideas are contagious: once they have been understood, they can be copied from mind to mind through language. In fact, they often seem to copy themselves without any decision on the part of their new host.
- Ideas are subject to independent discovery: an idea can be "in the air" and be expressed by a number of people. The fact that someone is the first to express an idea can seem morally arbitrary.
- Ideas are infinitely divisible with regard to utility value. If I share a good idea with you, I still have it.
- The only possible way to confine an idea is by keeping it secret. But even that cannot prevent someone from inventing it independently.

These claims are relatively straightforward. The real work comes when we attempt to determine whether or not the *works* created with these raw materials also inherit their unownable character.

The Works Created from Ideas by Nature Cannot and Should Not Be *Exclusively* Owned

Arguing for the common nature of works is more difficult than arguing for the common nature of ideas. In large part this is due to the dual nature of works: they are composed both of ideas and physical "expressions." Though some works (like a television broadcast) lack clear physical boundaries, many are expressed as physical objects that can be subject to exclusive ownership.

The only *universal* argument for the common status of works would be an argument against private property generally. As this route is blocked to us, we will need to take a more indirect approach. First, we will argue that it is appropriate to divide works into owned and unowned parts. Then we will argue that the *natural process* of creation creates moral obligations to share. Finally, we will appeal to the general principle that anything that can be shared without loss should be so shared.

All Works Can Be Divided Into Ownable and Unownable Features

No work is literally created *ex nihilo*: it requires some pre-existing physical substrate. This physical substrate is also the peg on which the individual identity of
the work can be hung. But every work also requires some pre-existing *conceptual* substrate. It must have enough resemblance to other works (and other
physical entities) that we can recognize it as a work and categorize it. Tolkien
would never have succeeded in sharing the Ring story if he had written it *entirely* in the "elfish" language he invented. In the same way, stories depend on what
IP lawyers call *scenes au faire*: stock situations and characters.

An author may very well have some natural property right in the total package
of a work, but it would be grossly unfair to assume that they own every detail of
their works. Such an assumption would be the equivalent to the way that Governor Winthrop "created" farmland at the Indians' expense. It would also be futile
and perverse to insist that an author attempt to identify the origin of every single
common idea in his work. As Jessica Litman convincing argues, no amount of
evidence about how a work was created could possibly prove its originality.[8]

In copyright law, the division between the owned and the unownable features
of a work are usually identified with the distinction between "idea" and "expression." As we have seen, both Alfred Yen[9] and Justin Hughes[10] argue that the
idea–expression distinction is grounded in natural law. Ideas become expressed
in works through the labor of authors, and it is this expressive labor that justifies
the special rights of authors.

Since New Works Are Constructed Out of Old Works, Authors Incur a Debt in Creating Works That Can Only Be Paid by Sharing Their Work

There is a kind of ecology associated with the process of creating works. Each
work contains new elements built on a foundation of old elements. Some of the
new elements will be associated with the expression of a work and contribute to
its unique character. Should authors have an absolute property right to the
unique features of their work? If so, then an author owes compensation to all of
his sources. Alfred Yen argues that the only way to avoid a vicious regress of
obligations is by forcing authors to "balance the books" by sharing their original
work. His argument runs as follows:

> Since the identification of one prior author as a person to be compensated
> merely raises the problem of identifying more persons, it quickly becomes clear
> that practically every author would both owe and be owed compensation under
> a complete property rights scheme. This implies that society could more fairly
> "balance the books" among all authors by simply recognizing the fact that au-
> thors will always owe a great deal to each other and letting it go at that. In other

words, society should forgive many of the "debts" owed by modern authors to their predecessors. In return for this windfall, modern authors should forgive similar debts to future authors. The effect of such a scheme would be to place even original material into a public domain from which future authors could borrow and to which they must contribute.[11]

The mechanism that guarantees this sharing with the future is limitation of term: at some point in the future all aspects of a work will enter the public domain.

Yen's argument is reminiscent of one Locke makes in the *First Treatise of Government*. When discussing filial obligation, Locke argues

> For to the Grand Father, there is due a long Score of Care and Expenses laid out on the Breeding and Education of his Son, which one would think in Justice ought to be paid. But that having been done in Obedience to the same Law, whereby he received Nourishment and Education from his own Parents, this score of education received from a Man's Father, is paid by taking care, and providing for his own children.[12]

At this point we must take care to distinguish two different arguments implicit in the quotes above. Yen's first point is that the assumption that authors have absolute property rights immediately puts them into an infinite regress of debts to predecessors. We examined arguments of this type in chapter 3 and found them plausible. However, Yen's proposed solution to the problem involves arguing that we can escape the regress by redefining the debts as debts to our successors (even as Locke suggests that fathers repay their parental debts by caring for their own children). This approach fails, for a number of reasons:

- It is inherent in the concept of "debt" that it involves a specific obligation between specific parties. While a creditor can choose to reassign the payment owed them to others, the *debtor* has no right to unilaterally reassign the debt.
- The object of Yen's arguments is to replace a specific debt with a general obligation. But the *intrinsic specificity* of debts makes it impossible to derive general obligations from them.
- The specificity of debts also entails that debts can be, in a sense, completed: when this debt is repaid, all moral obligations associated with it cease to apply. But general obligations are intrinsically open-ended. Does the fact that my parents raised five children mean that each of us only has the moral obligation to raise *one* child? If my parents were neglectful and parental obligations are a debt, why should my children expect more?

His argument may also run afoul of Rousseau's barbarians: "It was in vain to say 'I built this wall, I earned this land with my labor.' Who gave you your standing,

it might be answered, and what right have you to demand payment of us for doing what we never asked you to do?"[13] It is a general ethical principle that no one
is obligated to pay for a benefit they did not ask for. A greedy author would certainly be free to assert that he didn't *ask* that the intellectual commons be made
available to him and thus he has incurred no debt from using it. Such a misfit
would still be obligated to share the generic features of their work which we
identified above, and might also be subject to the kinds of arguments made
against those who want to opt out of morality.

Anything That Can Be Shared Without Loss Should Be/Must Be Shared With Anyone

Our final premise is the Grotian claim that anything that can be shared without
loss should be (weak version) or must be (strong version) shared. The weak version makes sharing superogatory and does not seem to require any detailed argument. It is the strong version that is the most interesting. It is strong enough to
require some sharing of works even without the claim that all authors are in debt
to their predecessors.[14]

Grotius argues for the strong version on the basis of natural law: since people
are sociable by nature, it is a "violation of the laws of society" to refuse an opportunity to do good without cost. It could also be argued that failure to do good
to someone in need (especially without cost) is morally blameworthy because it
represents an injury to the needy party.

Where We Have Gotten To So Far

The arguments so far have made a case for understanding ideas as things that are
intrinsically shareable and not ownable. This sounds an awful lot like a commons. We have also argued that the nature of ideas requires a moral framework
in which exclusive property rights are limited and some forms of sharing are
mandated. Finally, we have argued that the objects in this domain have a sort of
vitality: a life of their own. If we can establish that every person has a right to access this domain and use its resources we will have made a good case for building an IP theory on the concept of an intellectual commons.

IP and Natural Law

We now have the machinery we need to build a theory of intellectual property
that combines a natural law ethic with a theory of the commons. Our basic ap-

proach is to develop a metaphor for the economy and ecology of intellectual property based on the idea of renewable and intrinsically sharable resources. My name for this economy is "the intellectual commons." We will begin by further unpacking the geographic metaphors associated with the commons concept. We will then examine the differences between a natural commons and the intellectual commons. We will discover that in many ways the intellectual commons is a better model of common resources than a natural commons could be.

It should surprise no one that so much of the theory of property is shot through with geographic metaphors. Land was the ultimate holding through much of human history, and those habits of mind die hard. Even today, any land I own is my "real" estate, my "real" property. Every human being moves in and through a vast honeycomb of overlapping and distinct property rights. Some are barriers that must be worked around: others are bridges and paths that connect me to the resources of others. They create a "moral space" with a well defined topography. They also create an arena in which property can be created and transformed.

In what follows we will focus on three aspects of the land that will support useful analogies. Many others are possible, but these will motivate our discussion of the intellectual commons.

The earth provides *givenness*. Its existence necessarily precedes the existence and projects of any individual person. While people are not simple reflections of the land, it must be there. The idea of a "self-made man" is not only arrogant, it is literally incoherent. Second, the earth provides *material* and *tools*. The unowned things of the world are the ultimate foundation of ownership. Finally, the Earth is *inevitably common because it is not naturally infinite*. Anything physical I appropriate or despoil cannot be appropriated by someone else. Latecomers can only appropriate what others have not already appropriated. Though the land can replenish itself, it has a "carrying capacity" that cannot be exceeded. In a bounded world, all appropriation has moral significance.

Givenness

The commons shares the given, background character of the world. Every person finds the intellectual commons waiting for them when they are born, and it will survive the death of any individual person. We encounter language and the structures of culture in the same way that we encounter the constraints and regularities created by the laws of nature.

Independence from Human Projects

It might seem that the only way to argue that the intellectual commons is independent from human projects in the way that the physical world is would be by
arguing for the *metaphysical reality* of ideas and works. Perhaps the intellectual
commons is a platonic universe of archetypes waiting to be instantiated. If this
were true, then talk about "artistic creativity" would in a very profound sense be
misguided. The task of authorship would be a matter of realizing the idea in
some more-or-less adequate physical form. Most importantly, there would be no
intrinsic relationship between a work and an author that would support a claim
of ownership. If the author owns anything, it is the unique features of her "expression" of the idea.

   Fortunately, we need not be driven to such ontological extremes. The intellectual commons is ultimately independent of any *particular* human projects, since
it is not possible for an author to fully specify all possible interpretations or uses
for his work. As Arendt pointed out, our works (especially our ideas) can
achieve a kind of independence from us as they become so shop-worn that their
connection to a particular author fades away.

The Intellectual Commons Is the Source of Materials for
Constructing Intellectual Property

Our critique of the romantic theory of authorship made a case that all works depend on other works. It is also clear that any attempt to identify a set of *people*
who provided the materials for a work is ultimately doomed to failure.[15] Ultimately the raw materials for works emerge from the process of gleaning the
available stock of public cultural knowledge.

Human Laws and Social Institutions Can Create Situations
Analogous to Scarcity and Pollution in the Intellectual Commons

Though the commons has the transcendent characteristics listed above, it is still
something that must be accessed through human labor and cooperation. Restrictions on access are always a real possibility. Such restrictions can take many
forms: secrecy, denying access to physically realized works, and aggressively enforced property rights. Without some minimal level of continuous availability,
large sections of the intellectual commons can simply disappear: there are thousands of ancient works that only survive as titles.

   It is also possible to *try* to enclose parts of the intellectual commons through
technology or through coercive regimes of IP rights. All the radios in North Ko-
rea are "hardwired" to pick up a single radio station. The technology of trusted

system can turn a work from a thing to an "access event." Patent trolls can intimidate inventors with their restless rent-seeking. All of these acts of sabotage point to an even greater constraint: *the intellectual commons becomes useless when it cannot grow.* Its health requires more than the creation of works: it also requires that existing works continue to be available as the foundation for new works.

In the next section, we will take these analogies and try to extend them by bringing in ideas from environmental ethics.

Caring for the Commons: Analogies with Environmental Ethics

We have established that the intellectual commons has a givenness, vitality, and richness that transcends our individual histories. We have also established that in some extremely important ways it is not subject to some of the limits associated with a natural commons. This might suggest that the intellectual commons does not require our care the way that the natural commons does.

The last half-century has taught us (the hard way) that even something as vast as the Earth's atmosphere can be damaged and depleted by human activity. This is a lesson that we badly need to learn about the intellectual commons as well.

The independence of the intellectual commons cannot be understood as the independence of a platonic heaven. Remember that everything in the intellectual commons was made by people, and that unencumbered access to the resources of the intellectual commons occurs through the mediation of social and legal institutions. If this creative activity is hindered and institutional mediation fails, both the commons and the people who depend on it will suffer. In this section we will explore some analogies between the intellectual commons and the natural environment.

Environmental ethics is important because of a recognition of the finitude of nature and the effect that human activity can have on it. We can no longer appeal to the vastness of the world to argue that human acts can be evaluated in a purely local or individualistic way. Human beings have become so numerous and so powerful that their activities inevitably affect the natural world and the other people that live in it.

Environmental ethics seems to require the development of some positive conception of the common physical world. If the resources of the world are finite, then any act of consumption has consequences for other people and for future generations.

Dynamics: the Endless Need for Growth

Authors build their works from materials drawn from the intellectual commons. But if IP rights were permanent, nothing would ever flow back into the commons. It would not grow over time, and would become less and less useful as a starting point. Aspiring authors would face a tragedy of the anticommons: the cost of getting the rights needed to create a work would simply become greater and greater. Eventually the system would collapse.

Before authors can create, they have to have a starting place. This starting place is both a set of competencies with language and media and socially mediated knowledge and attitudes about the world. No work can be totally unconnected with earlier works, for the same reason that a truly "private" language would be no language at all. Thus the intellectual commons does provide tools and materials for the creative process.

There is also a sense in which any work that is shared publicly extends the intellectual commons, providing more material for later authors to exploit. The bottom line is this: no author creates *ex nihilo*, just as no individual exists who has not benefited from the care of others. The central question now becomes: do these connections create moral obligations?

There are many striking dualities in the concept of an intellectual commons. All forms of the commons require care to be sustained. A physical commons, like a pasture, requires limitations on use that enable it to sustain its carrying capacity. We must limit output to sustain the resource. In the intellectual commons, we need to promote input so that its carrying capacity will remain steady or grow. The central realities of a physical commons are boundaries and scarcity: the struggle is to prevent it from getting smaller. The central reality of the intellectual commons is that it can only be sustained through growth: the struggle is to promote the processes of creation and appropriation that sustain it.

Because the creation of new works requires an endless supply of new ideas, it is possible for legal and political barriers to create situations of scarcity. Scarcity and massive redistribution of wealth can also follow the introduction of new forms of IP, if care is not taken to maintain a just distribution of resources. The commons can also experience depletion from the extension of term and "propertizing" previously common ideas and works. IP protection can lock the developing world into the position of an eternal latecomer, always trapped in an inferior position by the scope and term of patents from the developed world.

Pathologies in the intellectual commons create economic and political pathologies. But it is also important to remember that the commons is the arena where nations have the greatest incentives to share and cooperate. Anything that undermines the health of the commons also undermines the health of the human community.

Scarcity: Privatization of Basic Research

One of the most disturbing trends affecting the intellectual commons is the increasing commercialization of basic research. The institutional context of patents has also begun to change. Congress passed a series of laws that allow federally-funded universities, government agencies, non-profit organizations, and federal employees to patent their inventions. The laws also provide tax incentives for private corporations to fund research and allow the creation of "cooperative research and development agreements" (public–private partnerships). The distinction between "pure" or "basic" research, funded by the government, and "applied" or "downstream" research pursued by private companies for commercial advantage is being erased. The net result is a process of privatization that steadily shrinks the domain of public knowledge.

The growing privatization of research has two kinds of adverse consequences. It undermines the norms of information–sharing that are a basic part of scientific research, and creates an anticommons that impedes both commercial and academic research. The decline of public research also exacerbates the problems of market failure: it further reduces incentives to search for treatments for diseases of the poor. It also changes the culture of basic science: the Technology Transfer Act [passed in 1986] allows researchers in government facilities, including scientists at the NIH, to patent their inventions and keep up to $150,000 of the yearly royalties on top of their government salaries. "The attitude today is exactly opposite of what I experienced in 1976," notes [Leonard] Hayflick. Now, "if you do not hold a patent on a cell population, plasmid, or microorganism, or if you are not a stockholder or scientific advisor to a company that exploits such materials, you are a failure in biology."[16]

Private researchers have strong incentives to avoid sharing information, in order to protect the potential patents on their research. They also have a strong incentive to "follow the money": to press research in the direction of funding rather than toward human needs or intellectual curiosity. The loss of cooperation alone introduces "friction" and inefficiencies in the course of research: this can only get worse as the profit motive increasingly drives research. A basic life-support system for the intellectual commons is being choked off, with predictable consequences.

Enclosure: DeCODE and the Genome of Iceland

The expansion of copyright term and the plethora of new patent types (on organisms, genes, business methods, algorithms, and a host of other things), along with the "right of publicity" that allows a person to treat any of their distinctive mannerisms as property, have created a new collection of economic rights that enrich a few at the expense of the many. Things that were once available free become more and more expensive as they are burdened with IP taxes. Faced with an increasingly restive customer base, IP owners are pressing for greater and greater control of consumer electronics and ever-more-sophisticated methods of control and surveillance. The dynamics of this spiral resembles early stages of land enclosure. As owners use sheer size to assimilate smaller businesses and dispossess the poor, they encounter increasing resistance from those with less and less to lose. The growth of IP rights has less and less to do with promoting any kind of social good.

In a sense, all IP rights are enclosures: all restrict access or use of some resource from the public. However, some projects seek restrictions so broad in scope that they literally "wall off" information from the intellectual commons. The efforts of deCODE genetics to achieve commercial control of the entire genome of Iceland is a case in point.

One of the most powerful weapons available to genomics is combined health records and genealogies for a large but relatively homogeneous population. The geneological/health data can be used to identify families with hereditary risk factors for a disease: DNA data from the family can then be used to search for the location of genes involved in the disorder. This technique was used to identify genes associated with cystic fibrosis, Huntington's disease, and early-onset Alzheimer's disease.[17] The motherlode of such data is the country of Iceland. Its population is big enough for large studies, but so homogeneous that most diseases will reflect single mutations. Researchers in Iceland have traced virtually every case of breast cancer to a man named Einar, who lived four hundred years ago. Einar had a mutation that produced a gene now known as BRCA2. Finding BRCA1 had taken over twenty years; BRCA2 was found in only two years because of the genetic homogeneity of the Icelandic population. In addition to this homogeneity, Iceland has documented geneological records for 75 percent of the population, some of which go back over 1000 years.

In 1996, Icelandic geneticist Kari Stepansson founded deCODE genetics to develop and exploit this genomic goldmine. Stepansson was able to convince Icelandic politicians to give deCODE exclusive rights to the genetic information of the entire population of Iceland. He was also given permission to combine the detailed medical records of the entire population with deCODE's DNA and geneological databases.

The deCODE project has been extremely controversial among genetic researchers and ethicists. Other researchers resent the need to pay deCODE to get at data they have generally collected from volunteers. The deCODE database represents the most intrusive documentation of health information ever collected, though deCODE claims that all personal data is independently encrypted before being entered into the database. Perhaps most troubling of all are questions of consent: the law setting up the database allows people to opt out, but the default is to include personal data. Thus thousands of Icelanders are having the most intimate details of their health histories collected and sold without their knowledge or consent. Iceland is poised to become the world's first gene–exporting country: "Some estimates put deCODE's potential value at more that twice Iceland's gross domestic product."[18]

DeCODE's activities are popular in Iceland, in large part because deCODE requires that any pharmaceutical company that develops a drug with deCODE data must make it available for free to the entire population of Iceland. It has been less popular elsewhere; in an editorial in the *New York Times*, geneticist R. C. Lewontin charged that "The conversion of the health and genetic status of the entire population into a tool for the profit of a single enterprise has been marked by the blatant political cynicism that is usually ascribed, with typical snobbery, to Mexico and Indonesia."[19] DeCODE is a purely private, for-profit company that is unlikely to damage the value of its franchise by disclosing any data through academic or non-profit channels.

The Conquistador Problem: Introducing New IP Systems

The dependence of the patent system on documented prior art makes it vulnerable to enormous "land grabs" whenever a new type of patent is introduced. In the absence of prior art, patent examiners sometimes grant enormously broad patent claims (such as "a method for updating the data of a computer through a communications channel") that then hamper the entry of other claimants into the market. Failure to control "conquistador" claims can lead to the stillbirth of an entire technology. The problem is that the patent system has no way to avoid granting very broad claims for a brand new type of patent. The patents that were intended to safeguard the development of technology ends up stifling it by creating an anticommons.

The conquistador problem is the inevitable consequence of a system designed primarily to *create* and *defend* property rights. It is also the consequence of treating patentable subject matter as the creation of private property from public resources. There is no equivalent to an "environmental impact statement" for the creation of new IP rights that takes into account the impact of new IP claims on users and on the availability of the commons itself.

Patents and the Developing World: The Problem of Latecomers

Since the IP system awards right to the first arrivals, it has created an endless race between claimants for IP rights. But the race is not being run on a level course: the developed world has a scientific and engineering infrastructure that the developing world will never be able to catch up with. Consider the dilemma of a country like Uganda, with at least a third of its population HIV-positive and a per-capita annual GDP (not income) of $1,700. At prevailing prices, a year's supply of the AIDS "cocktail" costs $15,000 per patient. Simple arithmetic says that Uganda needs to simply "write off" its HIV-positive citizens. Yet the same people would live if they had had the good fortune to be born in America or Europe. The law of necessity would say that if Uganda can manufacture its own AIDS drugs at a tolerable price, it has the right (even the duty) to do so, even if that means ignoring the IP rights of pharmaceutical companies. Though it may yet exercise that option, there is no way it will ever be able to "catch up" with drug companies enough to invent its own. The West will always have a twenty-year head start, thanks to the duration of patents.

This is a version of the problem of latecomers discussed in the previous chapter. To the extent that patents protect not only the fruits of research but the instruments of research, it becomes virtually impossible for a late entrant to ever compete.

All of the "environmental" problems described above reflect the way that the intellectual commons can act as both a gateway and a bottleneck to vital knowledge. They also illustrate the need to move in IP policy from an almost exclusive focus on individual claims to some kind of global perspective. The IP system could perhaps make the kind of structural changes that governments have made to seriously address issues like energy and pollution. The success of environmental laws in the developed world shows the possibility of making global changes in a complex legal and economic system without destroying it through mismanagement. Perhaps using environmental metaphors will help in the development of "IP management systems" that are both more just and more economically efficient than those currently in place.

A System of Intellectual Property Rights Based on the Intellectual Commons

The intellectual commons approach is most naturally specified as a pluralistic theory of IP rights. It provides a unifying context in which we can posit several sets of rights and trust in the ecology of the model itself to harmonize them.

The system of IP rights sketched below is an attempt to adjust the present IP regime to bring it in line with the theory of the intellectual commons. The approach is comprehensive and yet not radical: it does not posit new forms of IP or challenge the primacy of individualist models of property rights. Nevertheless, it does have some new features:

- The existence of a robust intellectual commons will be an explicit goal of the regime. The intellectual commons will become the kind of morally considerable entity that "the environment" is in environmental law.
- We will explicitly state rights associated with *using* IP. Current IP law only specifies a global social goal (the progress of science and the useful arts), and then attempts to implement that goal by specifying rights

for authors and owners of IP. My goal here is to curtail the expansion of IP rights by default: any new grants must be justified against the explicit baseline of user rights.

- We will mandate a kind of "environmental impact" process for introducing changes in the nature of IP: it will be necessary to explicitly justify changes by evaluating their likely effect on those with existing IP rights (which will now include users).
- We will strengthen the connections between the doctrine of compulsory licensing and the concept of eminent domain. Some particular works may have such overwhelming social value that it is appropriate for them to be considered public property. Compensation of the author may be appropriate in exchange for such a taking.
- The primacy of social good means that it will be possible for IP rights to *contract* as well as expand. In particular, terms may need to be made shorter (not just longer). Some classes of works may need to be "depropertized." Rights that create perverse incentives must be reassessed. Contraction of IP rights would not be considered a taking that required compensation, since they would apply to all IP owners equally. Socially destructive use of property (e.g., pollution, creating public hazards) is not a right.
- We will make an explicit connection between IP law and privacy law. However, this connection is not made by asserting that rights to privacy just *are* property rights. Rather, IP regimes must be constrained by the recognition of privacy rights. The connection between IP rights and privacy rights is a functional one: both involve the social regulation of information.

The system of rights given here is only one possible set of rules for managing common intellectual property. Much more radical approaches are possible, including new forms of IP (we will consider some of these possibilities in chapter 12). It is specific enough to lead to some practical debate and perhaps play a role in the ongoing development of IP rights.

I have chosen to posit three sets of rights and one global set of duties and obligations. This global set is intended to specify the overall social utility of the entire IP system. It provides the basic conditions for satisfying the overall function of the whole system (such as "the advancement of science and the useful arts"). The global duties and obligations then prescribe institutions and rules that will promote the rights of each group. As I argued in chapter 2, each group fulfills a necessary role in maintaining the health of the intellectual commons. I will assume in what follows that the rights listed below are lexically ordered.

Global Duties and Obligations

The purpose of an IP regime is to support the development and health of a system of access to common knowledge and the cultural treasures of human society. This access is subject to the constraints created by the rights of authors, users, and publishers.

1. The highest priority of an intellectual property system is the creation of a robust common store of human knowledge that is maximally available to all users, through promotion of the rights listed below.
2. The second priority of an intellectual property system is the creation of systems of property rights and economic incentives needed to meet priority (1). These rights and incentives will periodically be adjusted to meet priority (1).
3. The third priority of an intellectual property system is to *preserve* the existence and availability of works and ideas subject to the constraints of priorities (1) and (2).
4. Certain classes of knowledge can be designated as exempt from economic control. All persons have the unrestricted right to use or publish such knowledge.
5. Certain classes of knowledge can be designated as private. All access (including economic control) of private information must be constrained by ethical norms of individual dignity and autonomy.
6. All property rights associated with the intellectual property system must have limited term. Length of term must be inversely proportional to the restrictiveness of the property rights (i.e., the more restrictive the rights, the shorter the term). Length of term should also be inversely proportional to the rate of change in a body of knowledge (i.e., the faster technical change is occurring, the shorter the term).
7. All intellectual property rights associated with an individual are terminated not later than twenty years after the individual's death.
8. All intellectual property rights owned by corporate entities (other than trade secrets) must have a term not to exceed twenty years.
9. All particular grants of intellectual property rights can be challenged on the grounds of necessity or social utility (i.e., copyrights and other IP rights can be challenged and revoked the way patents can be).
10. All works must be made available in an unencumbered way at or before the expiration of term. Automated systems for rights management must disable themselves at or before the expiration of term.

The Rights of Users

Users are the foundation of the intellectual commons, since all authors and publishers must continue to be users. User's rights, unsurprisingly, center on use rights.[20]

1. Users have an unrestricted right to use intellectual property for any purpose except vending copies for economic gain, and to copy and re-organize that intellectual property as is necessary to facilitate this goal.
2. Users have an unconditional right to read anonymously.
3. User have a right to access intellectual property that is no longer available through market mechanisms.
4. Users have the right to use common stores of intellectual property (libraries) for any purpose other than vending works for profit.
5. Users have the right to become authors consistent with the other rights enumerated in this list.

The Rights of Authors

The economic rights of authors are always conditional on the health of the commons and the needs of users. However, authors have an inalienable "moral right" to have their authorship of a work acknowledged. I am unsure whether or not they have a right to preserve the integrity of their work. They must have first access to any economic benefits from their work in order to counterbalance the institutional power of publishers and the media.

1. Authors have an unconditional right to have their authorship of works acknowledged. This right is inalienable and is permanent.
2. Authors have the first and residual right to benefit economically from their works. This right can be assigned.
3. Authors have the right to preserve the integrity of their work against attempts to change its form or content. This right is inalienable and persists throughout the author's lifetime.
4. Authors have the right to control the economic uses of their work for a limited time. This right can be assigned.
5. Authors have the right to exclusively control the production and distribution of their work for a limited time. This right can be assigned.
6. Authors have the right to use existing works as the basis for new works, subject to the rights of the owners of the existing works.

The Rights of Publishers

Publishers must have economic rights necessary to make their dissemination of NEW works profitable within the constraints created by the rights of users and authors. They have a moral obligation to preserve the "independent thingness" of works by publishing them in a form that allows them to have an independent existence. Publishers must NEVER completely control access to works.

1. Publishers have the right to exclusively manufacture and sell original copies of a work that has been assigned to them and is still under IP protection, and to use that exclusive right for economic gain.
2. Publishers have no other intellectual property rights.

Comparison with Existing IP Law

This proposed system of rights accepts the legitimacy of IP as a system of property rights that originate with authors and which can be alienated as part of an economic system. The system of rights also continues to assume the "eternal triad" of authors, users, and publishers. However, the *balance* of rights is significantly shifted.

Unlike existing IP regimes, the intellectual commons regime involves an explicit set of criteria for maintaining the global health of the IP system. All economic rights associated with the intellectual commons must be periodically revised in order to guarantee that the regime continues to meet its stated goals.

The intellectual commons regime also explicitly assigns rights for users. The use rights granted here go far beyond the doctrine of fair use: in particular, they *guarantee* the right of users to access works that are no longer part of the market, something not done by the existing law. It specifically rejects the *NII White Paper*'s argument that fair use is a legal defense rather than an affirmative right. It also specifies that all grants of IP are subject to review and possible revocation.

The intellectual commons regime restores an author's moral rights in their work (something long rejected by American copyright law). The regime also assumes that authors are also users: the broad rights of users thus expand the ability of authors to use old works as material for new works.

Following this model would involve many specific changes in existing IP regimes. Infringement would become more difficult to prove, since the intellectual commons approach broadens the definition of acceptable borrowing. In most cases, the term of IP protection would be shorter. This is not expropriation from authors, since the same policy gives them much greater access to using the works of others, as Yen argued above.

The existence of the greater intellectual commons also exposes authors to the risk that their work will be deemed so socially valuable that they will not be allowed to assert property rights. This has the potential to discourage private investment in creating works of great social utility. We can try to mitigate this disincentive in two ways: by making it hard for IP rights to be denied (the current approach), or by accepting that societies are morally obligated to support basic research (the intellectual commons approach). Only the latter approach can be viable in the long term, since allowing the creation of many property rights creates the disincentives associated with an anticommons.

Conclusions

We have presented a theory of the intellectual commons that is based on an analogy between such a commons and the physical environment. This analogy is based on the nature of ideas and works, particularly their ability to transcend individual human projects. In the next chapter, we will apply the idea of an intellectual commons to the one form of IP most closely tied to utility: patents.

Notes

1. Justin Hughes, "The Philosophy of Intellectual Property," *Georgetown Law Journal* 77 (1988): 287.
2. Yen, "Restoring the Natural Law," 535-536.
3. Hannah Arendt, *The Human Condition* (Chicago: University of Chicago Press, 1958).
4. Arendt, *The Human Condition*, 50.
5. Arendt, *The Human Condition*, 52-53.
6. Arendt, *The Human Condition*, 55.
7. Foucault, "What is an Author?"
8. Litman, "The Public Domain," 1000-1004.
9. Yen, "Restoring the Natural Law," 537-538.
10. Hughes, "Philosophy of Intellectual Property."
11. Yen, "Restoring the Natural Law," 557.
12. Locke, *Two Treatises*, I. 91.
13. Jean-Jacques Rousseau, "A Discourse on the Origins of Inequality," 354.
14. The strong version is still weaker than an obligation to charity, since that might require sacrificial giving.
15. Litman, "The Public Domain," 1010-1011. "But when the author mines the raw material for her next work, significant portions of it will be the stuff of the outside world mediated by her experience. It is unsurprising, then, that parts of her work will echo the work of others."
16. Lori Andrews and Dorothy Nelkin, *Body Bazaar: The Market for Human Tissue in the Biotechnology Age*, (Westminister MD: Crown Publishing, 2001) 47-48.
17. Kevin Davies, *Cracking the Genome: Inside the Race to Unlock Human DNA* (New York: Free Press, 2001), 130-140.
18. Davies, *Cracking the Genome*, 136.

19. R.C. Lewontin, "People are not commodities," Editorial in *The New York Times*, January 23, 1999.

20. We are assuming that there is some reasonable way to distinguish between using works and publishing them: for more discussion on this point, see Jessica Litman "The Exclusive Right to Read."

8

The Commons in History

My interest in IP theory was aroused by James Boyle's book *Shamans, Software and Spleens*, which focuses on the challenges to IP theory created by new technologies such as cloning and software development. However, my attempt to understand Boyle's appeal to the idea of an intellectual commons quickly led me back in political history to the period between about 1500 and 1800.

This headlong flight into the past had three motivations. Intellectual property as a legal regime begins in late fifteenth- and early sixteenth-century Europe. If we want to rethink it, we would do well to find out what the originators of the system intended to do with it.[1]

This period in European history involved a transition from medieval to modern economic relationships, accompanied by a slow, yet vast transformation of property rights.[2] If this transition was correctly characterized as one of privatization, then understanding the system of property rights that existed before the change should tell me something about a system of non-private yet non-state property rights. We should also be aware of the role of the English Revolution in fomenting radical questioning of both the old order and the emerging order. One of the prominent themes of modern Locke exegesis is trying to understand his relationship to the radical wing of the English revolution.[3]

My final motivation was more theoretical (or at least more philosophical). If there ever was a golden age of philosophizing about property, it was during the emergence of modern political philosophy starting with Hugo Grotius and modern natural rights theories.[4] Grotius' early work *On the Freedom of the Seas* and chapter 5 of Locke's *Second Treatise on Government* provide the starting point for attempting to understand the not only the process of appropriation (transforming things into property), but also the significance of appropriation's necessary starting point: a common, unowned world. Grotius was motivated by the desire to legitimize free trade with the "new world" that had been revealed to Europe by the age of exploration, while Locke's theory was an attempt to show property rights come naturally "from below," instead of being imposed from above by an absolutist sovereign.

We will look at six historical contexts. The first is the thirteenth-century debate on apostolic poverty. This debate is important because it sharpened distinctions between ownership and "right to use," and because the critics of the Franciscan position introduced an interesting argument for interpreting individual property as part of nature. The second is Hugo Grotius' argument for freedom of navigation in his 1609 work *On the Freedom of the Seas*. Here Grotius introduces two arguments that will provide the basis of our analysis of the intellectual commons in the next chapter. The third is the enclosure movement, the long process of abolishing common property rights in English land.

After introducing enclosure, we will look at the ideological use of the commons concept in justifying the European seizure of land in the new world. We will then examine the attempt by Gerrard Winstanley and the diggers to "opt out" of a regime of private property entirely. Though Winstanley rejected expropriation, his critique of private property seems to lead in the same general direction as that of the European colonists in America. Finally, we will look at today's "true Levellers": the free software movement and their unruly fellow-travelers, the hackers.[5]

The Imagined Commons of Ancient Philosophy

Many ancient philosophers saw wealth as problematic. They imagined a golden age in which there was no private property and therefore no inequality or need. As Ovid described it, "No law or force was needed: men did right freely; without duress they kept their word . . . while the earth herself, Untouched by spade or plowshare, freely gave, as of her own volition, all men needed." As men left the golden age and declined through silver, bronze, and finally iron, "the ground itself, which had once been common to all, like sunlight and the air, Fell under the surveyor's drawn out lines."[6]

In the present, a ruler's wealth made him a hostage to fortune. Plato's solution in *The Republic* is have his ruling class have no private property. In advocating this approach, he was in part following the example of Lysurgas, author of the Spartan Constitution.

The Debate on Franciscan Poverty

The next precedent we will examine is the debate on Franciscan poverty. Francis of Assisi was determined that his brothers would live in "true apostolic poverty," which he took to mean that they would not claim ownership of anything. "The Franciscans claimed to have no property at all, either as individuals or as an order; and they claimed that in the things they used they had only simple 'use in fact' not use 'of right'; the things they used belonged, they said, either to the donor, or if the donor meant to give up all right in them, to the pope."[7] The theory of Franciscan poverty was developed by Bonaventure, Pope Nicholas III (a former protector of the order) in his Bull *Exiit* of 1279, Duns Scotus, and William of Ockham (among others).[8]

The basic tactic for defending the claim that Franciscans owned no property was to argue for a distinction between the use of a thing (*simplex usus facti*) and "dominion" over it. The Franciscans argued that all the property of the order was available to them for use, but that they did not have the right to alienate it or engage in financial transactions regarding it. Such an argument fails if *dominium* is intrinsic to exclusive use. The essence of the Franciscan position was the claim that common use is conceptually and ethically distinct from appropriation. Richard Tuck has summarized the position as follows:

> Common use, for Scotus, was not common *dominium*: it was not the case that the human race collectively had the kind of right over the world which (say) a Benedictine monastery had over its estates. Rather, each human being was simply able to take what he needed, and had no right to exclude another from what was necessary for him. But the crucial point is that such necessary use is not a case of *dominium utile*: Scotus took *dominium* to be necessarily private, something that could not only be exchanged, but which also could be defended against the claims of the needy, and quite possibly by violence. Such an account of man's natural life was absolutely in line with Franciscan poverty: in the terms of [the papal bull] *Exiit*, natural man had the *simplex usus facti*, but not *dominium*. As a corollary, the Franciscans in the early fourteenth century were supposed to be living a natural or innocent life (mirrored also, of course, in the poverty of the apostles).[9]

The controversy came to a head when Pope John XXII explicitly repudiated the Franciscan position in his bull *Quia vir reprobus* of 1329. Here John argued that property was natural to man and could not be avoided. At least some forms of common use involve the destruction of the fruits of the common (e.g., eating an apple). But if the fruit was destroyed *ipso facto* it becomes unavailable for the use of other commoners. We are thus faced with a dilemma: either argue that consuming use is illicit (thus rejecting God's providential care for his creatures), or we must accept that in the state of nature it is morally acceptable for a commoner to exercise exclusive use of a thing. But ownership (*dominium*) just means something like "the right to exclusive use of a thing." Thus private property is not only natural, but inevitable.

Ockham attempted to answer John's argument in his treatise *Opus nonaginta dierum*. He argued that the right to consume something is a necessary but not sufficient condition for *dominium*. Ockham asserts that *dominium* entails a much more expansive set of rights, including the right to seek legal sanctions, which are not inevitable consequences of consumptive use.[10] To really answer the argument, Ockham needs to give a positive argument to show that consuming use is not inconsistent with community. He attempts to make this case in a variety of ways and using a variety of illustrations. For example, consider a commoner who picks an apple in the commons and eats it. According to feudal law, that apple was already owned by the lord of the manor and also by God. Yet no one would argue that the man's destruction of the apple constituted an injustice. Therefore the mere consumption of and thus the destruction of *dominium* in the thing is not an injustice. Moreover, it does not require that the commoner assert an exclusive right to the apple. Ockham also appeals to an example similar to Cicero's example above. Consider the guests at a party. They have the right to consume the host's food and drink and possibly have a right to lodging. However, they do not *own* the host's food and drink: they have no right to empty his larder and take the food home with them.[11] In Cicero's analogy, the fact that I have a right to a seat in the theater does not give me the right to demand the same seat every time, or to pull the seat off the floor and take it home with me.

Whatever success Ockham might have achieved with these arguments was blunted by his acceptance of the claim that natural man had *iura* over the natural world. By making this concession, Ockham was forced to accept the idea that there could be individual claim rights in the state of nature. But accepting this claim was tantamount to saying that natural man had a *dominium* in things. All subsequent attempts by Ockham to definitionally split *ius* and *dominium* were ultimately irrelevant.

The failure of the Franciscan project meant that it became necessary for any theory of the commons to provide some explanation of the relationship between common rights and individual rights. Either the commons would have to be abandoned as an explanatory construct, or some way had to be found to make common rights and individual use fit together.

Natural law theorists found another promising approach: a distinction (which Pufendorf described as understandable "even by those of average mental discernment") between the commons and its fruits. "The oak tree was no one's, but the acorns that fell from it became his who had gathered them."[12] Commoners can have exclusive rights to the fruits of the commons while having inclusive rights in the commons itself. The distinction makes another approach possible as well: instead of merely focusing on what is *taken*, we can justify appropriation by focusing on what is *left*. Consumptive use is consistent with a commons as long as we leave "as much and as good" for other commoners.

Grotius: Freedom of Navigation

In 1603 a Dutch ship seized a Portuguese trading vessel in the Indian Ocean, in response to the Portuguese claim of an exclusive right to trade in the East Indies. When a Dutch court ruled that the seizure was legal, a number of Mennonite shareholders in the Dutch East India company objected, and the company retained Hugo Grotius to write a defense of the seizure. Grotius wrote a long treatise (*On the law of prize and booty*) in response. Only the twelfth chapter of this work (*On the freedom of the seas*) was published during Grotius' lifetime, in 1609.[13]

Grotius made two separate lines of attack on the Portuguese position. First, he argued that the Portuguese had no right to claim the countries of the East Indies, since they were already occupied. Second, he argued that since the sea has an un-limited ability to support navigation, the Portuguese had no right to impede any-one else's use of it. He summarizes his general position as follows:

> Two conclusions may be drawn from what so far has been said. The first is that which cannot be occupied, or which has never been occupied cannot be the property of anyone, because all property has arisen from occupation. The sec-ond is, that all which has been so constituted by nature that although serving some one person it still suffices for the common use of all other persons, is to-day and ought in perpetuity to remain in the same condition as when it was first created by nature.[14]

The first argument identifies a class of unownable (because unoccupiable) things, and shows that the sea naturally falls into this class. The second argument is *moral* rather than conceptual: Grotius is saying that infinitely sharable things *should not* be property, whether or not they can be occupied. Hear him thunder against the Portuguese:

> If in a thing so vast as the sea a man were to reserve to himself from general use nothing more than mere sovereignty, still he would be considered a seeker after unreasonable power. If a man were to enjoin other people from fishing, he would not escape the reproach of monstrous greed. But the man who even pre-vents navigation, a thing which means no loss to himself, what are we to say of him?

> If any person should prevent any other person from taking fire from his fire or a light from his torch, I should accuse him of violating the law of human society, because that is the essence of its very nature, as Ennius has said: "No less shines his, when he his friend's hath lit." *Why then, when it can be done with-out any prejudice to his own interests, will not one person share with another things which are useful to the recipient, and no loss to the giver?* These are ser-vices which the ancient philosophers thought ought to be rendered not only to foreigners but even rendered for nothing. But the same act which when private possessions are in question is jealousy can be nothing but cruelty when a com-mon possession is in question. For *it is most outrageous for you to appropriate a thing, which both by ordinance of nature and by common consent is as much mine as yours, so exclusively that you will not grant me a right of use in it which leaves it no less yours than it was before.*[15]

The sea is *de facto* common because it lies outside human power to occupy it. But Grotius wants to make a much stronger point, one which possesses direct ap-plication to IP. He is arguing that anything that can be shared without loss must be treated as common, *whether or not it can be occupied.* This obligation de-rives directly from "the law of human society," and thus appears to create a strong presumption in favor of common use whenever this is possible. Note also that, unlike the rule of necessity, this obligation to share derives from the nature of the entity alone, without regard for need.

Grotius believed that the commons were not restricted to the primordial state of nature: entities with the right nature (like the sea and fire) never leave the original community. It is also important to remember that when private property was instituted, the commoners kept a "benign reservation in favor of the primitive right" of common use. Thus, "in direst need the primitive right of use revives, as if community of ownership had remained."[16]

Grotius sees three reasons that a resource should be treated as a commons. It may have always been treated as common ("common consent"), or it may be common by an "ordinance of nature": it may be impossible to enclose or it may be something that can be shared without loss.

The Enclosure Movement and the Decline of Open-field Agriculture

The commons was more than an abstraction in seventeenth-century England: it was still a real, though waning, reality.[17] In most parts of England, the "common" fields were owned by the lords at the sufferance of the king, an arrangement dating back at least to the Norman conquest. The least fertile land (the "waste") was used for pasturage and gathering firewood, while fertile lands were used for pasture and also for cultivation. Social norms restricted the use of commons to common and renewable uses. Commoners were not allowed to build houses on common land, or to build fences or plant hedges. Commoners were also forbidden to use the "capital" of common lands: it was forbidden for them to cut down trees or to mine the earth. Generally, cultivated land was divided into unfenced strips: a typical farmer would own several strips scattered across the fields. The scattered and unfenced nature of each peasant's holdings created a strong pressure toward uniform agricultural practices.

The commons were a significant economic resource for most peasants and an absolute necessity for those without a trade or their own holdings. The forests and wastelands also provided a "liberated zone" for those who wished to live outside the prevailing economic system. "Unlike the relatively stable and docile populations of open arable areas, these men, cliff-hanging in semi-legal insecurity, often had no lords to whom they owed dependence or from whom they could hope for protection."[18]

The commons were a reflection of pre-modern land tenure. These systems were defined by the existence of "customary" rents, a fixed set of social and economic relationships, and overlapping sets of property rights that made sale of land almost impossible. The operative paradigm was rent rather than ownership: each proprietor of a piece of land had duties and rights with respect to it. Since no one proprietor owned any piece of land, none had the authority to make an investment and expect a private return for that investment.

The common fields were threatened by ingrossing (replacing common human use with pasturage for a landlord's animals) and enclosure (dividing common fields into restricted, private units of production). Ingrossing directly displaced commoners: according to Thomas More, they "leave no ground for tillage, they inclose all into pastures; they throw down houses; they pluck down towns and leave nothing standing but only the church to be made into a sheep fold . . . They turn all dwelling places and all glebe land into desolation and wilderness." [19]

Enclosure was a significantly more complex process than the brute expropriation of ingrossing. It involved the conversion of customary rent relationships to fee simple rent relationships based on cash. The conversion to cash forced out poorer tenants, opening the way for the remaining tenants to consolidate their holdings and enclose them with fences or hedges. The final stage was to privatize common land through an act of parliament. The common land was put up for sale: since only the gentry could afford to buy it, it rapidly became consolidated as the private property of the wealthy. From this point the new owners were free to develop it or to exploit it through logging and mining.

The poor who lost their common rights were forced to leave the countryside or become wage laborers, since they now had to pay cash for food, fuel, and lodging. For them, enclosure was a deprivation of traditional economic rights and social relationships. It was a tremendous boon for the new landowners, since they were now freed from traditional restrictions on land use. Those who supported enclosure argued that it would increase the efficiency of agriculture, and the available evidence supports this claim. Those who opposed it saw it as dispossession and the rejection of community in favor of individual greed and inequality.[20] The process both supported and depended on the emergence of a purely private and exclusive system of land ownership, and the reduction of all land tenure to either "private" or "public" (managed by the state). The third alternative disappeared.

Enclosing America: Natural Law Arguments for Expropriating Indian Land

In *Utopia*, More argued that in his ideal world, failure to use resources justified expropriation. The Utopians deal with overpopulation by sending colonists to found new cities on the nearby continent:

> And if the population throughout the entire island exceeds the quota, then they enroll citizens out of every city and plant a colony under their own law on the mainland near them, wherever the natives have plenty of unoccupied and uncultivated land. Those natives who want to live with the Utopian settlers are taken in. When such a merger occurs, the two peoples gradually and easily blend together, sharing the same way of life and customs, much to the advantage of both. For by their policies the Utopians make the land yield an abun-

dance for all, which previously seemed too barren and paltry even to support the natives. But if the natives will not join in living under their laws, the Utopians drive them out of the land which they claim for themselves, and if they resist make war on them. The Utopians say it's perfectly justifiable to make war on people who leave their land idle and waste, yet forbid the use of it by others who, by the law of nature, ought to be supported from it.[21]

The Utopians have two moral justifications for their acts of colonization. The first is their need. The second is their productivity: the Utopians are so good at agriculture that they can make the land support both themselves and the original inhabitants. To deny them access to unused or "under-used" land, by Utopian standards, is to do them harm and justify war.

James Tully sees a strong connection between Locke's theory of property and English colonialism in the New World.[22] According to Tully,

> Locke defines political society in such a way that Amerindian government does not qualify as a legitimate form of political society. Rather, it is construed as a historically less developed form of European political organization located in the later stages of the "state of nature" and thus not on a par with modern European political formations. Second, Locke defines property in such a way that Amerindian customary land use is not a legitimate form of property. Rather, it is construed as as individual labour-based possession and assimilated to an earlier stage of European development in the state of nature, and thus not on an equal footing with European property.[23]

Tully illustrates these points by describing a dispute between John Winthrop and Roger Williams in 1633. "Williams argued that the royal patent did not convey title to Indian land and that the only legitimate means of possession was by treaty with Amerindian nations in order to acquire rights of usufruct on their property, as he did in Rhode Island and the Dutch in New York. Governor Winthrop replied that the Indians possessed only what they cultivated; the rest was open for appropriation without consent."[24] The following is a pastiche of arguments for Winthrop's position:

> "We did not conceive that it is a just Title to so vast a continent, to make no other improvements to millions of acres in it, but onley to burne it up for a pastime," [John] Cotton rejoined. "As for the natiues in New England," Winthrop explained, "they inclose noe Land, neither have any settled habytation, nor any tame cattle to improue the land by, and soe have no other but a Naturall Right [i.e. in the products of their labor]." "The Indians" [Francis] Higginson concurs, have no right in their traditional lands because they "are not able to make use of the one-fourth part of the Land, neither have they any settled places . . . nor any ground as they challenge as their own possession, but change their habitation from place to place." Since they possess very little land, America is *vacuum domicilium*, a "vacant" or "waste" land and so *Vacuum Domicilium cedit occupanti*. "In a vacant soyle," Cotton points out to Williams, "he that taketh possession of it and bestoweth culture and husbandry upon it, his right it

is." Enunciating a principle similar to Locke's famous proviso in section 27, Winthrop concludes "soe . . . if we leave them [the Amerindians] sufficient for their use, we may lawfully take the rest, there being more than enough for them and us."[25]

Note that in this context, Winthrop is not appealing to any justification in terms of necessity. There is no reason for him to, since America is still in a state of nature. Since the Indians are so wasteful, the European seizure of their land still leaves "more than enough for them and us."

This whole edifice of expropriation depended on the belief that America was still in a state of nature. This belief in turn was justified by the belief that only those who enclosed and "improved" land had the right to assert ownership of it. This belief in turn was sustained by a willful refusal to see non-exclusive use as a right. It is not difficult to see that the same forces that were driving enclosure in England were also driving a more aggressive form of development in the colonies.

The Diggers

Given the arguments about waste and expropriation given above, it is perhaps unsurprising that someone would apply the same logic to England itself. That person was Gerrard Winstanley, who led an unsuccessful attempt to create a new communal society on common land. The diggers began their colony in April 1649. Winstanley and his "diggers" were driven off the common land in less than a year: in March 1650 their homes and fields were burned, and they were forcibly dispersed. It is unlikely that the colony ever numbered more than a few dozen people.

That the diggers are remembered at all is due to the torrent of pamphlets Winstanley wrote between 1648 and 1650 (totaling almost 700 pages in Sabine's edition). Winstanley's arguments are couched in the language of religious enthusiasm, but they rest on a finely honed sense of natural rights and social justice. His message to "those who call yourselves lords of manors and lords of the land" was simple and direct:

> [T]he earth was not made purposely for you to be the lords of it, and we to be your slaves, servants and beggars; but it was made to be a common livelihood to all, without respect of persons; and that your buying and selling of land and the fruits of it, one to another, is a cursed thing.[26]

Despite the ferocity of this rhetoric, Winstanley was not a violent revolutionary. "England is not a free people until the poor that have no land have a free allowance to dig and labour in the commons, and so live as comfortably as the landlords in their enclosures."[27] The diggers intended to return at least the common parts of the earth to its original moral state:

> If you look through the earth, you shall see that the landlords, teachers, and rulers are oppressors, murderers and thieves in this manner. But it was not thus from the beginning. And this is one reason of our digging and labouring the earth, one with another, that we might work in righteousness and lift up the creation from bondage. For so long as we own landlords in this corrupt settlement, we cannot work in righteousness; for we should still lift up the curse and tread down the creation, dishonour the spirit of universal liberty, and the work of restoration . . . and that not only this common or heath should be taken in and manured by the people, but all the commons and waste ground in England and in the whole world shall be taken in by the people in righteousness, not owning any property; but taking the earth to be a common treasury, as it was first made for all.[28]

Winstanley rejected as "Norman enslaving entanglements" the rules forbidding plowing and sowing in common land: "One third part lies waste and barren, and her children starve for want, in regard the lords of manors will not suffer the poor to manure it."[29] The diggers believed that their occupation of common land did no harm to either landlords (since they were not expropriating the landlord's holdings) or to the common people, since for all intents and purposes they *were* the common people.

The diggers basic argument went like this:

1. The earth was originally given to mankind by God ("Reason," in Winstanley's writings) in common so that individuals can sustain themselves through their labor.
2. Since the earth was given in common, no one has the right to appropriate the land itself and exclude others from its use.
3. All people have the natural right to use the resources of the earth in order to labor and sustain themselves. The rich have already exercised this right through the "theft" of establishing private property.
4. The poor have a *prima facie* right to use land not owned or used by anyone in order to sustain themselves. This right extends as far as their ability to labor on the land extends.
5. This right is a right of use (a *dominium utile*), not a right of appropriation. The commons can never be enclosed or sold.
6. The customary restrictions of usage of the commons were a plot by landlords to keep the commons undeveloped and thus force the poor to work as wage laborers.
7. No one in a state of surplus has the right to forbid a person in want from using the earth to meet their survival needs.

Principles (2) and (3) undermine existing property relationships, even though Winstanley was unwilling to directly challenge them. Thus it is unsurprising that the wealthy opposed the diggers. It is also not surprising that most commoners also opposed them, since they were advocating the use of commons in a way that would change it irreversibly.

It is instructive to consider the differences between Winstanley's arguments and the arguments given by Winthrop and his defenders in the previous section. Ideologically, Winstanley would be closer to the Native Americans than to the settlers. The settlers are attempting to argue that they have a right to appropriate land from a commons without consent since the land is being mismanaged by its present inhabitants. Winstanley is arguing that the poor have the right to develop the commons because no one has the right to appropriate land and exclude others from its use. The settlers are treating the property of others as if it were *res nullius*; Winstanley is treating the commons as if they were truly *res communis*.

There is one point at which both parties seem to agree: that traditional common modes of use do not endow a commoner with any natural property right. In their own way, the diggers are as determined to improve the commons as the enclosers were: if their enthusiasm for "manuring" and other modern farming practices was typical, it probably explains why enclosure occurred largely without violence. It is a misunderstanding to see the diggers as part of an atavistic reaction to economic change: their communism and concern for optimum use of resources are thoroughly modern.

A Contemporary Parallel: The Free Software Movement and Bulgarian Virus Writers

In the *Gnu Manifesto*, Richard Stallman explains his decision to turn his back on the world of proprietary software and begin writing a complete operating system and tool chain from scratch.[30] His basic reason was a desire to opt out of a property system that seemed increasingly corrosive to social goods:

> I consider that the golden rule requires that if I like a program I must share it with other people who like it. Software sellers want to divide the users and conquer them, making each user agree not to share with others. I refuse to break solidarity with other users in this way. I cannot in good conscience sign a nondisclosure agreement or a software license agreement. For years I worked within the Artificial Intelligence Lab to resist such tendencies and other inhospitalities, but eventually they had gone too far: I could not remain in an institution where such things are done for me against my will. So that I can continue to use computers without dishonor, I have decided to put together a sufficient body of free software so that I will be able to get along without any software that is not free. I have resigned from the AI lab to deny MIT any legal excuse to prevent me from giving GNU away.

Part of the manifesto is a kind of catechism for free software. Here are two of the questions:

"Don't programmers deserve a reward for their creativity?"

If anything deserves a reward, it is social contribution. Creativity can be a social contribution, but only in so far as society is free to use the results. If programmers deserve to be rewarded for creating innovative programs, by the same token they deserve to be punished if they restrict the use of these programs.

"Shouldn't a programmer be able to ask for a reward for his creativity?"

There is nothing wrong with wanting pay for work, or seeking to maximize one's income, as long as one does not use means that are destructive. But the means customary in the field of software today are based on destruction. Extracting money from users of a program by restricting their use of it is destructive because the restrictions reduce the amount and the ways that the program can be used. This reduces the amount of wealth that humanity derives from the program. When there is a deliberate choice to restrict, the harmful consequences are deliberate destruction. The reason a good citizen does not use such destructive means to become wealthier is that, if everyone did so, we would all become poorer from the mutual destructiveness. This is Kantian ethics; or, the Golden Rule. Since I do not like the consequences that result if everyone hoards information, I am required to consider it wrong for one to do so. Specifically, the desire to be rewarded for one's creativity does not justify depriving the world in general of all or part of that creativity.

Stallman's programs and his arguments are reminiscent of Winstanley's: like the Diggers, Stallman is attempting to build an alternative society from scratch. And, like the Diggers, he is refusing to expropriate the resources that he considers held by oppressors. In this effort he is aided by the nature of software, since it is virtually the only valuable thing that can be built almost entirely from labor. Unlike the Diggers, Stallman's efforts have not been in vain. The free software movement has now produced over twenty million lines of high-quality software, and it has become possible to work entirely in a free software environment.[31]

In the passage above, Stallman denounces restrictions on program use as "deliberate destruction." This destruction is manifested both as wasted labor and as corrosion of the social contract. Stallman clearly sees social good as being at least equal to or trumping individual good: "Specifically, the desire to be rewarded for one's creativity does not justify depriving the world in general of all or part of that creativity." In effect, Stallman is arguing that sharing socially useful works is obligatory.

Stallman sees no need to expropriate, but his argument puts expropriation within reach. If you have a duty to share your software with me, I have a right to help myself to it.

In "The Bulgarian and Soviet Virus Factories" Bulgarian computer scientist Vessilin Bontchev documents the rise of the computer virus subculture in Bulgaria.[32] Many virus writers got their start stealing access to PCs that sat unused in plant manager's offices:

> Since the introduction of computers in the Bulgarian offices was not a natural process, but due to an administrative order, very often these computers were not used—they were only considered an object of prestige. Very often on the desk of a company director, near the phone, stood a personal computer. The director himself almost never uses the computer—however sometimes his/her children came to the office to use it—to play games or to investigate its internals.[33]

The most obvious reason for the rise of virus writing was the government's deliberate decision to create an outlaw computer culture: "At that time most Western software was copy protected. Instead of training our skilled people in writing their own programs, we began to train them to break copy protection schemes. And they achieved great success in this field."[34]

It is not hard to imagine the frustration that a Bulgarian hacker might feel at watching a computer sit unused day after day. Shouldn't things be available to those who want to do something with them? In a few simple steps, the hacker is Winthrop and the director's computer is nothing but a "unused" land waiting to be exploited.

Does Labor Trump Occupation?

The examples given above show the risks associated with one commoner deciding whether or not another was "really" using the commons. The moral hazards of such self-interested judgments are obvious. Grotius and others believed that property rights could be overridden by the rule of necessity. However, there seems to be no compelling moral reason to expropriate someone else's share of the commons (or their private property) in order that it be used more intensively. The greed of Winthrop and his followers blinded them to the fact that the Indians were using their land in a sustainable way, and that there really was no great surplus of unused resources. The anger of the Bulgarian hackers blinded them to the fact that they were simply replacing one irrational monument to selfishness with another. The only absolute right to labor is the right to labor for survival. Any claim of exclusionary rights based on lesser reasons simply uses the commons as a cover for theft.

The Lessons of History

The episodes considered above teach us a number of important lessons about the commons. First, the existence of commons is compatible with private ownership:

essentially consumptive use shows that there is a clear sense in which private property is natural. It is also clear that common ownership is not an idea restricted to either the state of nature or "pre-modern" political systems. Some resources so totally outstrip the human ability to enclose that they will always be understood in some way or other as common. It is also possible that there are resources that can be enclosed, but *should not* be, since sharing them in no way detracts from their utility value.

Governor Winthrop's arguments for expropriation and Winstanley's communism show us the risks of assuming that "common" means the same thing as "underutilized according to me." In effect, Winthrop is claiming the right to determine who is a commoner on the basis of their use of common resources. But this is to effectively destroy the commonness of the resource. His pious claims of protestant virtue are simply a device to privatize a commons by force. Appropriation without consent does not take place in *the commons* unless it can occur without violating the rights of other commoners.

The development of property rights from St. Francis to Richard Stallman has definitely moved in the direction of privatization: private property has become the default mode of ownership. At the same time, Francis, Winstanley, and Stallman all bear witness to a continuing undercurrent of common-property thinking. All three tried (and Stallman is still trying!) to build a common world in the margins and interstices of the world of private property. Though their efforts have been defeated again and again, the continual resurgence of a common vision is good evidence that understanding property must involve understanding property that is not private.

Notes

1. L. Ray Patterson, *Copyright in Historical Perspective*, (Nashville, TN: Vanderbilt University Press, 1968).

2. This transformation is discussed in terms of economic theory in Douglass North and Robert Thomas, *The Rise of the Western World: A New Economic History* (New York: Cambridge University Press, 1973) and in terms of historical texture by Fernand Braudel, *Civilization and Capitalism*, vol. II, *The Wheels of Commerce* (New York: Harper and Row, 1979), 247–296.

3. See Richard Ashcraft, *Revolutionary Politics and Locke's Two treatises of Government*. Ashcraft's theories are controversial: see also Sreenivasan, *Limits on Lockean Rights*, 13–17, 42–43.

4. Buckle, *Natural Law and the Theory of Property*.

5. Purists should note that a more appropriate term would be "crackers" or "vandals." The free software movement still uses the term "hacker" as a compliment: a classic hacker is a highly skilled and tremendously focused programmer.

6. Ovid, *Metamorphoses*, book one, lines 89–100. Quoted in Thomas More *Utopia*, (Norton Critical Editions. New York: Norton, 1975. Translated and edited by Robert Merrihew Adams), 96.

7. See John Kilcullen, "The Origin of Property: Ockham, Grotius, Pufendorf, and

Some Others" www.humanities.mq.edu.au/Ockham/wpr.html.

   8. Tuck, *Natural Rights Theories*, 20–25.

   9. Tuck, *Natural Rights Theories*, 21–22.

   10. Tuck characterizes this argument as "desperate and unconvincing," an attempt to refute John by simply redefining the term under dispute. I am inclined to be more charitable.

   11. Kilcullen, "The Origins of Property," 3.

   12. Pufendorf, *Political Writings of Samuel Pufendorf*, 185.

   13. J. B. Schneewind, *The Invention of Autonomy: A History of Modern Moral Philosophy* (Cambridge: Cambridge University Press, 1998), 70.

   14. Grotius, *On the Freedom of The Seas*, 23–24.

   15. Grotius, *On the Freedom of The Seas*, 45. emphasis added.

   16. Grotius, DJBP, 193.

   17. B. A. Holderness, *Pre–industrial England: Economy and Society from 1500 to 1750* (Lanham, MD: Rowman and Littlefield, 1976).

   18. Hill, *The World Turned Upside Down*, 43.

   19. More, *Utopia*, 14.

   20. Karl Marx, *Capital*, Volume 50, *Great Books of the Western World* (Chicago: Encyclopedia Britannica, 1991), 354–368.

   21. More, *Utopia*, 45.

   22. James Tully, "Rediscovering America: The *Two Treatises* and Aboriginal rights" Chapter 5 of *An Approach to Political Philosophy: Locke in Contexts.* (New York: Cambridge University Press, 1993), 137–176.

   23. Tully, "Rediscovering America," 138–139.

   24. Tully, "Rediscovering America," 148–149.

   25. Tully, "Rediscovering America," 150–151.

   26. Christopher Hill, *Winstanley: The Law of Freedom and Other Writings*, (Cambridge: Cambridge University Press, 1973), 99.

   27. Hill, *Winstanley*, 87.

   28. Hill, *Winstanley*, 87–88.

   29. Hill, *Winstanley*, 115.

   30. Richard Stallman, "The GNU Manifesto" www.gnu.org.

   31. Eben Moglen, "Anarchism Triumphant: Free Software and the Death of Copyright." *First Monday* 4, no. 8 (1999) emoglen.law.columbia.edu.

   32. Vessilin Bontchev, "The Bulgarian and Soviet Virus Factories," 1991, www.texfiles.com/virus/bulgfact.txt (accessed April 12, 2004).

   33. Bontchev, "The Bulgarian Virus Factories."

   34. Bontchev, "The Bulgarian Virus Factories."

9

Social Utility and the Rise of
the Imperial Author

Although sometimes referred to as "rights" of the users of copyrighted works, "fair use" and other exemptions from infringement liability are actually limitations on the rights of copyright owners. Thus, as a technical matter, users are not granted affirmative "rights" under the Copyright Act; rather, copyright owner's rights are limited by exempting certain uses from liability. –NII White Paper, 84.

Copyright began to assume its modern form in England with passage of the Statute of Anne in 1710. For the first time, copyright became a right of authors rather than publishers. The Statute of Anne is also notable for shifting the rationale of copyright from censorship to the encouragement of learning. L. Ray Patterson argues that the statute was intended as "trade regulation": he sees the primary purpose of the statute as breaking the monopoly on printing held by the Stationer's Company.[1]

A more orthodox interpretation sees the statute as the first clear acknowledgment that "copyright is about the conditions of authorship." I favor a third interpretation: the Statute of Anne was essentially a rejection of the Stationer's monopoly and a return to the tradition of patents. Patents have always been a right of "authors" (inventors), and have always had a rationale based on the advancement of knowledge. This interpretation is supported by the decision to give copyrights that same fourteen-year term as patents.

The last four centuries has seen a gradual shift of emphasis (at least in copyright) in the relative importance legislatures and judges have given these two criteria. I will examine the beginnings of the shift toward author's rights, and I will examine the end stage of this process: the almost complete absorption of social value into authors' rights: in effect, the claim that what is good for authors is automatically good for society as a whole. We will also explore the impact of broadcast and recording media on the interpretation of IP rights; in particular, a shift from interpreting works as things to works as events. This shift is highly significant, since access to events has a different dynamic: instead of buying a thing like a book, one buys the right for a single performance. The concept of trusted systems, a technology that allows publishers to completely control and commodify every use of a work, represents an enormous transfer of power away from users. In the final section we discuss whether or not IP rights should be socially mandated or individually contracted.

Valuational Solipsism

It might appear that libertarianism offers the least constrained framework for relating ethics and polity, but even here there may be implicit constraints determined by an individualistic bias. Phillip Pettit identifies two attitudes toward individuals implicit in analytic political philosophy.[2]

The first of these he calls "personalism," summed up in Jeremy Bentham's famous dictum "Individual interests are the only real interests. Take care of individuals: never injure them, or suffer them to be injured and you will have done well enough for the public."[3] Pettit sees personalism as a "plausible and harmless working assumption." However, he also discerns another assumption in analytic political philosophy that he describes as having a "massive, warping" effect on political theory: the assumption he calls valuational solipsism. He describes it as follows:

> The assumption of valuational solipsism is the assumption that any property that can serve as an ultimate political value, any property that can be regarded as a yardstick of political assessment, has to be capable of instantiation by the socially isolated person, by the solitary individual. It is the assumption that the ultimate criteria of political judgment—the reserve funds of political debate— are provided by non-social as distinct from social values.[4]

Pettit sees valuational solipsism as depending on another assumption, which he calls social atomism: "The social atomist holds that the solitary individual—the agent who is and always has been isolated from others—is nevertheless capable, in principle, of displaying all distinctive human capacities."[5] Pettit identifies social atomism as the distinctive feature of Anglo-American political theory: he sees the opposite position (which he calls holism) as a distinctive feature of European political theory, starting with Rousseau.

Pettit sees atomism as a problem because it pushes the entire evaluative framework away from any any criterion that takes social utility into consideration: "What the reasoning establishes is that if the isolated individual is allowed to figure among the alternatives to be evaluated in political theory, as it has traditionally figured, then atomism filters social values out of the set of ultimate political criteria; it drives the theorist to countenance only non-social values as the ultimate terms of political assessment."[6] Atomism stacks the deck against any account that assumes that humans are inherently social beings whose humanity can only be expressed in social relationships. At best, social utility is a second-order concept that can only be justified in terms of some kind of social contract that asocial individuals accept because it promotes individual utility.

Property Rights

The influence of atomism is reflected in the modern theory of rights. Ronald Dworkin's definition of rights as individuated political aims makes this emphasis clear:

> A political right is an individuated political aim. An individual has a right to some opportunity or resource or liberty if it counts in favor of a political decision that the decision is likely to advance or protect the state of affairs in which he enjoys the right, even when no other political aim is served and some political aim is disserved thereby, and counts against that decision that it will retard or endanger that state of affairs, even when some other political aim is thereby served.

Social atomism sees the central task of political philosophy as that of constructing a social theory that is either derived directly from the psychology of self-interested individuals (e.g., Hobbes), or that imposes the minimum set of social norms required to force self-interested individuals to respect the rights of other individuals. As Jeremy Waldron puts it, "The case of fraternity illustrates the point that there may be some values (or concerns or commitments) which, because of their communal character, cannot be captured in the language of rights. This is what we should expect . . . The idea of rights emerged with the growth of ethical and political individualism, and we should expect that rights will be associated at a fundamental level with individualistic concerns." [7]

There are few areas of social theory where social atomism has exerted a stronger influence than in the theory of property. This can be seen plainly in William Blackstone's famous definition of property as "that sole and despotic dominion which one man claims and exercises over the external things of the world, in total exclusion to of the right of any other individual in the universe." [8]

Economics as Rationality

The history of IP law is a history driven by the history of technology. As new forms of work and new media emerge, IP law is forced to constantly reinvent itself. The constant need for extension to new domains forces an equally constant return to the foundations of IP law. [9]

Though property law and political theory provide the bones of such a framework, its muscle and sinew are provided by economic theory. IP law is in large part an attempt to create a framework of legally sanctioned economic incentives that will promote policy goals.

Geoffrey Brennan argues that modern economic theory has made two significant methodological contributions to political discourse. [10] The first is rejection of utilitarianism in favor of Paretan models. The second is "realism about virtue": if we prefer a self-organizing economic polity, we must favor one that somehow reconciles individual utility with social utility. Any system that ignores or downplays individual utility can only be maintained by coercion and central planning. There is also a deep sense of skepticism about whether efficient central planning is even possible, since it seems to require a Laplacean demon's level of knowledge about the effects of institutional forces.

The global insights described above are matched by an account of individual choice based on economic rationality. Amartya Sen characterizes this account of rationality as follows: "It is fair to say that there are two predominant methods of defining rationality in mainline economic theory. One is to see rationality as internal consistency of choice, and the other is to identify rationality with maximization of self-interest."[11]

Sen criticizes the latter on grounds similar to objections Pettit makes to social atomism: it rules out entire classes of motivations by fiat.

> The self-interest view of rationality involves *inter alia* a firm rejection of the "ethics-related" view of motivation. Trying to do one's best to achieve what one would like achieve can be a part of rationality, and this can include the promotion of non-self-interested goals which we may value and wish to aim at. To see any departure from self-interest maximization as evidence of irrationality must imply a rejection of the role of ethics in actual decision taking (other than some variant of the exotic moral view known as "ethical egoism"). The methodological strategy of using the concept of rationality as an "intermediary" is particularly inappropriate in arriving at the proposition that actual behavior must be self-interest maximizing. Indeed, it may not be quite as absurd to argue that people always *actually* do maximize their self-interest, as it is to argue that *rationality* must invariably demand maximization of self-interest. Universal selfishness as *actuality* may well be false, but universal selfishness as a requirement of *rationality* is patently absurd.[12]

The absurdity Sen alludes to has two senses. First, it may well be in a purely self-interested agent's interest not to appear self-interested in order to manipulate others. However, the absurdity here is not merely a matter of tactics. An economic system presupposes a stable social order, and such an order is only made possible if we can assume that others will keep their commitments even when they are out of our sight. As Hobbes so eloquently showed, no such order naturally emerges out of the interactions of totally egotistical individuals.[13]

Ethical Shorthand: Reducing Social Utility to Author's Rights

Justification for creating or expanding author's rights has become so routine that it has collapsed into a kind of intellectual shorthand. As a typical example, consider the following argument from a recent European Union directive on intellectual property:

(10) If authors or performers are to continue their creative and artistic work, they have to receive an appropriate reward for the use of their work, as must producers in order to be able to finance this work. The investment required to produce products such as phonograms, films or multimedia products, and services such as "on-demand" services, is considerable. Adequate legal protection of intellectual property rights is necessary in order to guarantee the availability of such a reward and provide the opportunity for satisfactory returns on this investment.

(11) A rigorous, effective system for the protection of copyright and related rights is one of the main ways of ensuring that European cultural creativity and production receive the necessary resources and of safeguarding the independence and dignity of artistic creators and performers.[14]

This argument has a number of features worthy of note. First, note the emphasis in (10) on the "creative and artistic" nature of the work of authors and performers. The protections afforded by IP are apparently intended to safeguard "the independence and dignity" of these important people. In short, the purpose of IP law is *the protection of authors*, who deserve this treatment because of the special nature of their work. But what must be protected? The ability of authors to make a "satisfactory return on their investment" by selling their work.

The directive above wraps a layer of rhetoric about creativity around a hard core that reduces authorship to the production of a commodity. It is unusual in its offhand inclusion of publishers ("as must producers in order to finance this work"): most versions of this argument never even mention the people that make virtually all the money from IP. It is very typical in including no mention at all of the public or any public interest in IP. Arguments of this sort pass immediately from the claim

1. Society needs what artists and inventors produce (i.e., the work of authors has social utility)

to the claims

2. Society must create and protect economic incentives for artists and inventors.
3. The only economic incentives are property rights.

and finally to the claim

4. Works are the property of artists and inventors.

It is at this stage that the argument can rhetorically kick away reference to the needs of society and focus totally on the individual property rights of authors. After all, property rights are basic rights that must be protected by the state *sui generis*. Thus it is no surprise that the only rights that seem important in IP law are the rights of authors.

But can we really eliminate considerations of social utility? The focus on property rights also creates an almost irresistible temptation to replace (1) with something like

I.   People will pay for what artists and inventors produce (i.e., the work of authors has economic value)

This would enable us to eliminate social utility entirely. The problem is that (I) doesn't seem to imply (2) in the way that (1) does. In fact, it seems to render (2) pointless: there is no need to create economic incentives if they are already there!

All is not lost, however. We can replace (I) with

i.   Works are subject to a public goods problem: they are expensive to produce but are either usable by many people without depletion or are trivially cheap to copy.

(i) explains why government intervention might be necessary (since works can't be "locked up" like bicycles), but it still doesn't imply (2). Consider an entrepreneur who introduces "bottled air": a glass jar containing ordinary air. His product is certainly subject to a public goods problem, but it is by no means obvious why the state should step in to guarantee that he can make money.[15]

The problem with both (I) and (i) is that they beg the ultimate question: why should the state act to enforce the economic value of IP rights? The only really plausible reason always comes back to social utility: works are generally useful and important, and their (artificially created) economic value reflects some more general form of value.

There are other points of contention with the argument on the previous page as well. It embraces a point of view that Eben Moglen identifies with mythical creature called "the econodwarf," who believes that the only motivation of authors is economic gain. The problem is that the econodwarf's theory of human behavior is grossly inadequate: "But no matter how often he hears *Don Giovanni* it never occurs to him that Mozart's fate should, on his logic, have entirely discouraged Beethoven, or that we have *The Magic Flute* even though Mozart knew very well he wouldn't be paid."[16]

We can identify two sets of assumptions that might lead us to identify the whole rationale of the IP system with the rights of authors. The first is that there is a pre-established harmony between the rights of authors and the social utility of IP as an institution. Taking care of authors and their needs will automatically make the system produce social goods. The second is that providing economic incentives to authors is necessary to guarantee the production of works, and that the most efficient way to provide such incentives is by vesting authors with property rights.

These assumptions are intended to support consequentialist arguments for intellectual property rights couched in terms of social utility. But how is such social utility determined?

Social Utility as Literary, Scientific, or Practical Merit

The stated rationale for IP law in the United States is "promoting the progress of science and the useful arts." If this is the case, how do we tell what promotes such progress (and thus is worthy of IP protection) and what doesn't? The patent system requires prospective justification of a patent claim in terms of the invention's novelty, utility and non-obviousness. Thus the grant of a patent is directly supported by some plausible claims of (at least) the usefulness of the invention. There has never been a similar examination process associated with copyright. Instead, the first copyright statutes specified classes of works that were presumed to have social utility (published writings and maps). Both the Statute of Anne and the 1790 Copyright Act mandated the deposit of copies of works in central libraries: the fact that the Statute of Anne required deposit in several libraries gives good reason to assume that the purpose of such deposit was availability of the work for use.

The extension of the field of copyright in the last two hundred years can be understood as the system being extended in response to changes in technology: new kinds of work being found to deserve copyright protection. But this is only part of the story. Copyright law has now been extended to cover works that play no role whatsoever in "the progress of science and the useful arts," such as grocery lists and post-it notes.[17] There is still significant legal debate over whether or not commercial speech like cigarette advertisement is covered by the First Amendment, but there is no doubt as to whether or not it deserves IP protection. Why did this change occur?

The most sympathetic interpretation sees this expansion of scope as a democratizing process that rejected the snobbery that identifies social utility with scholarship and high art. In 1870 Congress extended copyright to "a painting, drawing. chromo, statue, statuary and . . . models or designs intended to be perfected as works of the fine arts."[18]

When Oliver Wendell Holmes Jr. extended copyright protection to advertising art in the 1903 *Bleistein v. Donaldson Lithographing Company* decision, he argued:

> Certainly works are not the less connected with the fine arts because their pictorial quality attracts the crowd and therefore gives them real use—if use means to increase trade and to help to make money. A picture is none the less a picture and none the less the subject of copyright that it is used for an advertisement. And if pictures may be used to advertise soap, or the theater, or monthly magazines, as they are, they may be used to advertise a circus . . . if they command the interest of any public, they have a commercial value—it would be bold to say that they have not an aesthetic and educational value—and the taste of the public should not be treated with contempt. It is an ultimate fact of the moment, whatever might be our hopes for a change. That these pictures had their worth and their success is sufficiently shown by the desire to reproduce them without regard to the plaintiff's rights.[19]

The circus posters deserved protection because of their economic value, but also for a much deeper reason:

> The [work] is the personal reaction of an individual upon nature. Personality always contains something unique. It expresses its singularity even in handwriting, and a very modest grade of art has in it something irreducible, that is one man's alone. That something he may copyright.[20]

Holmes seems to be arguing that any form of original expression deserves copyright protection without regard to any broader criteria of social value. And the definition of originality is broadened to include any expression of personality, since personality always contains something unique. Thus it is not hard to see how grocery lists end up being copyrighted. This broadening has the effect of relieving judges from the burden of value judgments, which Holmes considers a good idea:

> It would be a dangerous undertaking for persons trained only to the law to constitute themselves final judges of the worth of pictorial illustrations, outside of the narrowest and most obvious limits. At the one extreme some works of genius would be sure to miss appreciation. Their very novelty would make them repulsive until the public had learned the new language which their author spoke. It may be more than doubted, for instance whether the etchings of Goya or the paintings of Manet would have been sure of protection when seen for the first time. At the other end, copyright would be denied to pictures which appealed to a public less educated than the judge.[21]

Broadening the definition of fine art to this extent has the effect of rendering it trivial and ultimately irrelevant. Everything that isn't a bald-faced copy of some other work is "original," and on the basis of that originality alone is worthy of copyright protection.

Perhaps the last vestige of such a criterion is embedded in the 1976 Copyright Act's definition of fair use: "for purposes such as criticism, comment, news reporting, teaching (including multiple copies for classroom use), scholarship, or research" (17 U.S.C 107). After quoting this passage, the *NII White Paper* authors rather dismissively point out "The recitation of assorted uses . . . has been held neither to prevent a fair use analysis from being applied to other 'unlisted' uses nor to create a presumption that the listed uses are fair."[22]

If we make aesthetic or educational value irrelevant to decisions about copyright we have done more than clear the way for a purely market-driven economic analysis. We are also well on the way to removing the advancement of science and the useful arts as relevant criteria for evaluating IP policies. Unless there is some other kind of social utility to replace them, IP seems to simply reduce to economics.

The collapse of aesthetic or educational value as a criterion for the social utility of IP still leaves open other possible criteria for the way that IP can advance the sciences and the useful arts. The most obvious is using property rights to directly promote the growth of knowledge.

Promoting the Growth of Knowledge

The IP system could be interpreted as an attempt to create an economic value for the production and dissemination of knowledge. But how is this supposed to work? A great many works are created in order to achieve some non-economic goal. Writers write to express themselves, or out of a desire to change the world. Inventors invent in order to solve various kinds of practical problems, without necessarily intending to make the invention itself a product. Scientists study nature in search of understanding or in order to participate in the shared vision of the scientific culture.[23]

The fact that science, art, and invention managed to proceed for centuries without IP protection certainly suggests that it is not a necessary condition for the growth of knowledge. There is no point in providing people an incentive to do what they would already do anyway.

One could argue that IP, by giving economic value to "pure" knowledge, encourages its production. IP helps make the publication and dissemination of scholarly works economically viable, but it doesn't provide a way for most scientists and scholars to make a living the way a writer of romances or thrillers can (an unsurprising fact, given that most scholarly journals don't pay their authors or their referees).

At any rate, the "purest" knowledge (scholarship, basic scientific research) is precisely what we don't want to be private property. Ownership is just as likely to impede the progress of science as promote it. In their discussion of the impact of privatization on biomedical research, Heller and Eisenberg noted the following:

> In biomedical research, as in a post-socialist transition, privatization holds both promises and risks. Patents and other forms of intellectual property protection for upstream [i.e., basic research] discoveries may fortify incentives to undertake risky research projects and could result in a more equitable distribution of profits across all stages of R & D. But privatization can go astray when too many owners hold rights in previous discoveries that constitute obstacles to future research . . . The result has been a spiral of overlapping patent claims in the hands of different owners, reaching ever further upstream in the course of biomedical research. Researchers and their institutions may resent restrictions on access to the patented discoveries of others, yet nobody wants to be the last one left dedicating findings to the public domain.[24]

We will look in detail at patents and basic research in chapter 11. Here let us simply note that IP rights provide a powerful incentive for the production of works but no direct incentive for the production of knowledge per se. Creating or promoting a "market" in basic research helps solve one economic problem (public goods) by introducing the possibility of others (rent seeking and the possibility that specialized communities will be ignored as potentially unprofitable).[25]

Nevertheless, these arguments are the strongest basis for connecting social utility with publisher's rights and will be addressed further in chapter 12.

Creating a Market

In the preceding sections, we looked for a kind of material role for IP law in promoting specific kinds of social goods, such as fine art or the promotion of knowledge. There remains the possibility that IP law promotes social utility directly, not by encouraging any particular kind of work, but in simply creating a market for ideas. IP allows authors to control access to their work, thus creating economic value for it. Publishers can use the market to determine a work's ultimate value, and their investment in producing the work is protected by their monopoly rights. IP law should make creating and disclosing works more attractive, because it guarantees that an author who is willing to make her work public can count on it not being appropriated by others.

Edward Samuels has compared the market created by copyright to a lottery.[26] Authors "roll the dice" by publishing their works. A few will become extremely popular and thus extremely wealthy, though most will gain little or nothing. Samuels argues that the possibility of winning provides motivation for authors to share their work (though it may not be the reason they created them). In this view, the role of IP rights is not to *directly reward* the creation of social value, but to create a framework in which it becomes possible for market forces to reward the creation of social value.

It is not difficult to see how such an arrangement might promote economic utility, though it will often be grossly inefficient. Robert Frank and Phillip Cook have argued that the lottery metaphor creates "winner–take–all" markets that actually undermine a variety of social ideals:

> But winner take all markets also entail many negative consequences . . . Winner take all markets have increased the disparity between rich and poor. They have lured some of our most talented citizens into socially unproductive, sometimes even destructive, tasks. In an economy that already invests too little for the future, they have fostered wasteful patterns of investment and consumption. They have led indirectly to greater concentration of our most talented college students in a small set of elite institutions. They have made it more difficult for "late bloomers" to find a productive niche in life.[27]

The problems listed above can only be resolved if there are non-market-driven institutions supporting the growth of knowledge. Economic rights may have a role to play, but they cannot possibly be the whole story.

Social Utility and the Imperial Author

The real message of copyright expansion is that judgments about quality have no place in understanding copyright. The function of copyright is to guarantee that any work that seems valuable to someone can be vended through the market. Worries about quality represent a confusion of levels: we have the NEH and NEA and NIH to promote the production of social utility. The function of IP law is to create and maintain conditions under which the market can bring users together with works. Such a market can only exist if we legally enforce the property rights of authors. The social utility that an IP system promotes is simply a byproduct of "the magic of the marketplace."

Thus we arrive back at the EU argument on page 134. We need to have strong IP protections because they protect the economic interests of authors, and what is good for authors is good for everyone else. As new media and new kinds of works become possible, we must create systems that allow them to be treated as property. If property rights drive innovation, then more property rights will create even more innovation. If protecting the property status of digital works involves intrusive and aggressive interference with the behavior of users, then this is just the price that must be paid. The only relevant question we need ask about IP policy is: how can authors make money?

We are perilously close to a line of argument pointed out by the legal realist Felix Cohen:

> There was once a theory that the law of trade marks and trade names was an attempt to protect the consumer against the "passing off" of inferior goods under misleading labels. Increasingly the courts have departed from any such theory

and have come to view this branch of law as a protection of property rights in divers economically valuable sale devices. In practice, injunctive relief is being extended today to realms where no actual danger of confusion to the consumer is present, and this extension has been vigorously supported and encouraged by leading writers in the field. Conceivably this extension might be justified by a demonstration that privately controlled sales devices serve as a psychologic base for the power of business monopolies, and that such monopolies are socially valuable in modern civilization. But no such line of argument has ever been put forward by courts or scholars advocating increased legal protection of trade names and similar devices . . . Courts and scholars, therefore, have taken refuge in a vicious circle to which no obviously extra-legal facts can gain admittance. The current legal arguments runs: One who by the ingenuity of his advertising or the quality of his product has induced consumer responsiveness to a particular name, symbol, form of packaging, etc., *has thereby created a thing of value, a thing of value is property; the creator of property is entitled to protection against third parties who seek to deprive him of his property. . . . The vicious circle inherent in this reasoning is plain. It purports to base legal protection upon economic value, when, as a matter of actual fact, the economic value of a sales device depends upon the extent to which it will be legally protected* . . . The circularity of legal reasoning in the whole field of unfair competition is veiled by the "thingification" of property. [28]

We must protect IP because it has value. But Cohen's argument points out that we can't simply identify value and cost. Many valuable things don't come with a price tag on them. This could be due the fact that they are a brand-new thing that hasn't be given a market value yet. It could also be because they are public goods (valuable but not amenable to economic transactions), or because we have decided that making them property would be an unjust restriction on the rights of those who use them. We could make anything and everything property if we are willing to interfere enough in people's lives. But the fact that making something property gives it economic value does not always make that an appropriate thing to do. As Edwin Hettinger points out, every new property right involves giving one party an economic advantage by imposing economic costs on others. Unless the rights of those who must pay the cost are irrelevant, this means we must offer some justification for any decision to create a property right.[29]

The enormous expansion of copyright began in 1831, with a statute that specified that sheet music could be protected by copyright. This small and plausible extension set the stage for a completely new domain of IP: works that captured or "fixed" ideas rather than directly expressing them. The preceding analysis unpacked the moral shorthand that ignores social utility or identifies it with author's rights. We are now ready to look at the evolution of IP theories toward the "imperial author."

Conclusions

In this chapter we have seen the transition from a publisher-centered understanding of IP to the development of an author-centered understanding. We have also seen how the idea of "the advancement of science and the useful arts" has been reduced to something like "making a work publicly available." This has led to the moral shorthand that identifies social utility with guaranteeing the property rights of authors.

If authors are the only ones with rights, then IP regimes are no longer agreements designed to use property as an means of promoting social utility. They are property rights that are guarded by the state, and any limitation on them is an "unfair subsidy" or *noblesse oblige*. In the next chapter we will look at attempts to restore user rights and social utility as factors in IP regimes by attacking the imperial author.

Notes

1. Patterson and Lindberg, *The Nature of Copyright*, 28.
2. Phillip Pettit, "Analytical Philosophy," in Goodin and Pettit, eds., *A Companion to Contemporary Political Philosophy* (Oxford: Blackwells, 1995), 7–38.
3. Pettit, "Analytical Philosophy," 24.
4. Pettit, "Analytical Philosophy," 26.
5. Pettit, "Analytical Philosophy," 28.
6. Pettit, "Analytical Philosophy," 30.
7. Waldron, *The Right of Private Property*, 91.
8. William Blackstone, *Commentaries on the Laws of England*. Quoted in Rose, *Authors and Owners*, 7.
9. Goldstein, *Copyright's Highway*, 9.
10. Geoffrey Brennan, "Economics" in Goodin and Pettit, eds., *A Companion to Contemporary Political Philosophy*, 123–156.
11. Amartya Sen, *On Ethics and Economics* (Oxford: Basil Blackwell, 1987), 12.
12. Sen, *On Ethics and Economics*, 15–16.
13. Amartya Sen, "Rational Fools," *Philosophy and Public Affairs*, vol. 6 (1977): 314–344.
14. Excerpted from Directive 2001/29/EC of the European Parliament and of the Council of 22 May 2001, "On the Harmonization of Certain Aspects of Copyright and Related Rights in the Information Society."
15. The fact that bottled air is easy to make doesn't affect the argument. He could go through some incredibly expensive and pointless procedure to put air into jars without creating some reason that the state should guarantee his economic success.
16. Moglen, "Anarchism Triumphant."
17. Litman, "The Public Domain," 974.
18. Samuels, *Illustrated History of Copyright*, 140.
19. Goldstein, *Copyright's Highway*, 61.
20. Samuels, *Illustrated History of Copyright*, 129.
21. Samuels, *Illustrated History of Copyright*, 142.

22. *NII White Paper,* 75.

23. Robert Merges "Property Rights Theories and the Commons: The Case of Scientific Research," in *Scientific Innovation, Philosophy, and Public Policy,* edited by Ellen Paul, Fred Miller, and Jeffery Paul (New York: Cambridge University Press, 1996), 145–167.

24. Michael Heller and Rebecca Eisenberg, "Can Patents Deter Innovation?: The Anticommons in Biomedical Research," *Science* 280 (1998): 698–699.

25. Putting research under the discipline of the market creates pressure to demand payment for any kind of cooperation whatever (rent seeking); it also raises the possibility that the needs of those without substantial economic power will be ignored as potentially unprofitable (e.g., treatments for a disease that only affects five hundred people in the world, or a disease of the desperately poor). See appendix A for general discussions of rent–seeking and market failure.

26. Samuels, *Illustrated History of Copyright,* 16–17.

27. Robert Frank and Philip Cook, *The Winner–Take–All Society: Why the Few at the Top Get So Much More Than the Rest of Us,* (New York: Penguin, 1995), 4.

28. Felix Cohen, "Transcendental Nonsense and the Functional Approach," *Columbia Law Review* 35 (1935): 809, 814–817, emphasis added.

29. Edwin C. Hettinger, "Justifying Intellectual Property," *Philosophy and Public Affairs* 18, no. 1 (1989): 31–52.

10

The Author Metaphor

It is true that cyberspace vastly enhances the power of users of works: it not only makes works available to individual users more easily and in greater volume than in the hard copy world, but also communicates them in an easily manipulated format that users can instantly recopy, adapt or forward to other users. Nonetheless, the perspective of user rights, albeit important, should remain secondary. Without authors, there are no works to use.

—Jane Ginsburg

The publisher-centered theory constituted the basic framework of the earliest IP regimes: IP rights just were the rights of patent holders and groups like the Stationer's Company. To my knowledge such a theory has no modern advocates. User-centered theory seems never to have been given center stage. The most dominant and "naturally" monolithic theories are author-centered theories.

According to an author-centered theory (herein referred to as AC), IP exists to protect the property rights of authors from publishers (who want to pirate works) and users (who want to copy and use works without cost). Authors merit such property rights because either (a) they create their works *ex nihilo*, or because (b) they creatively transform their raw materials into something unique and new, or because (c) they deserve compensation for the labor they invest into making works. These arguments are important because they have always played a central role in efforts to expand IP rights, and because they have a natural resonance with the individualistic theories of property that have dominated western political thinking. They have also been used as ideological cover for attempts to promote publisher-centered IP regimes: it is a common rhetorical tactic for publishers to argue that certain changes in IP law should be opposed because they would harm authors.

What Is an Author?

As we saw in chapter 4, the copyright system as originally developed was a system of property rights for publishers, not authors. The eighteenth century witnessed a transformation of the ideology of copyright to its present form. How and why did this transformation take place?

We can find significant theoretical resources for addressing this question in the work of Michel Foucault, in particular his essay "What is an Author?"[1] In this paper, Foucault articulates a functional and hermeneutic account of author-ship. It is functional in that he identifies authorship with a varying set of social and legal roles, and hermeneutic in that Foucault sees the concept of authorship as both constitutive of discourse about works and also a critical theoretical con-struct of that discourse.

It seems clear that the concept of proprietor is not essential to the concept of an author, since authorship existed for thousands of years before the IP frame-work was developed. Foucault identifies two earlier functional aspects of author-ship, which we will call "forensic" and "warranting." Foucault reminds us that "In our culture (and doubtless in many others), discourse was not originally a product, a thing, a kind of goods; it was essentially an act—an act placed in the bipolar field of the sacred and the profane, the licit and the illicit, the religious and the blasphemous."[2] So one of the original functions of authorship was iden-tifying who should be punished when discourse transgressed social, legal, or moral norms. This forensic function was codified in the licensing acts regulating printing in sixteenth- and seventeenth-century England.

The second functional aspect of authorship sees attribution as a kind of war-rant or promise. Foucault argues that before the eighteenth century attribution was considered unimportant for what we would call "literary" works, but was considered critical for scientific and theological works:

> There was a time when the texts that we today call "literary" (narratives, sto-ries, epics, tragedies, comedies) were accepted, put into circulation and val-orized without any question of the identity of their author: their anonymity caused no difficulties since their ancientness, whether real or imagined, was re-garded as a sufficient guarantee of their status. On the other hand, those we would now call scientific—those dealing with cosmology and the heavens, medicine and illnesses, natural sciences and geography—were accepted in the Middle Ages as "true" only when marked with the name of their authors. "Hip-pocrates said," "Pliny recounts," were not really formulae of an article based on authority: they were markers inserted in discourses that were supposed to be re-ceived as statements of demonstrated truth.[3]

In effect, the attribution of authorship served as a sort of warrant for the truth or significance of a work. Foucault argues that the transition to a modern scientific world view has led to a reversal in the relative importance of authorship in litera-ture and science. The reputation of a particular scientist plays a significant socio-logical role in the development of science, but not an epistemic one: the merit of the work is determined by its methodology and its relationship to the rest of sci-entific knowledge. The connection between a scientific theory and a particular scientist is ultimately irrelevant to the scientific significance of the theory. "The author function faded away, and the inventor's name served only to christen a theory, proposition, particular effect, property, body, group of elements or patho-logical syndrome."[4]

The Romantic Model of Authorship

While authorship has become less important in science, Foucault argues it has become more and more important in literary theory. If we take the goal of literary theory to be something like "why is this text the way it is?" or "how can we characterize the relationship between these particular works?", then the author (as a real historical person with a real personality) becomes the ultimate explanatory construct:

> The author provides the basis for explaining not only the presence of certain events in a work, but also their transformations, distortions, and diverse modifications (through his biography, the determination of his individual perspective, the analysis of his social position, and the revelation of his basic design). The author is also the principle of a certain unity of writing—all differences having to be resolved, at least in part, by the principles of evolution, maturation, or influence. The author also serves to neutralize the contradictions that may emerge in a series of texts: there must be a certain level of his thought or desire, of his consciousness or unconscious—a point where contradictions are resolved, where incompatible elements are at last tied together or organized around a fundamental or originating contradiction.[5]

In short, the concrete and singular character of an author justifies explaining literary facts in terms of historical or psychological facts: the author is the bridge between literary theory and the rest of human knowledge.[6]

Accepting the author as theoretically central to literature also implies an account of literary works as in some sense an expression or outgrowth of a particular person's psychology and historical situation. If this is so, the relationship between a work and a particular author is not purely arbitrary as the relationship between (say) a scientist and a scientific theory. If Einstein hadn't discovered the theory of relativity, someone else would have. But if *War and Peace* is in some sense an expression of Tolstoy's unique situation and personality, there is a sense in which no one but Tolstoy could have written it.

The most striking difference between modern IP and full liberal ownership is limited term. As we saw in the preceding chapter, even the earliest patent systems limited the term of patent grants. In Honoré's typology, limited term is characteristic of more limited forms of ownership such as leases and licenses. A renter has an exclusive right to the use of the rental, but cannot sell it or permanently alter it without permission. This partial ownership is derivative from the property rights of the landlord. Is there a similar set of property rights underlying intellectual property?

Arguments for Author-centered Theories of IP

AC theorists argue that the act of creating a work invests the author with basic rights of ownership and control. We can distinguish between two lines of argument: an analogical argument that simply applies the rules of real property to intellectual works, and an argument from creativity.

Analogical Arguments

We must begin by distinguishing between the assertion of an analogy and some kind of justification for it. The analogy has a kind of surface plausibility, especially when we compare physical and intellectual works. The sandbox in my back yard is mine because (a) I acquired the material for it legally, (b) I built it on my property, and (c) I did all the work that went into making it. This *magnum opus* is mine because (a) I didn't plagiarize any of it, (b) I came up with the idea for it without help from anybody else, and (c) I did all the work that went into making it. This intuitive justification of IP rights turns on two kinds of justification: compliance to law (a), and labor/creation as justifying ownership (b and c). We could call the simple assertion of the analogy that presupposes a theory of real property the "analogical argument."

A strictly analogical account of IP ownership faces both practical and theoretical difficulties. The first difficulty is the fact that IP rights are prima facie different from real property rights.[7]

Copyright gives me some say in what you do with my book even after I sell it to a publisher and he sells it to you. When someone else buys my house, I fully alienate it: I give up the right to tell them what they can and can't do with the sandbox. There is a plain difference between stealing my shoes and pirating my book. If you steal my shoes, I can't use them any more: you now have the power to exercise "despotic dominion" over them. In the case of piracy, I can still use my book. I can protect my shoes from theft with locks and bars, but there is no way to exercise that kind of control over a published work (at least not yet: see chapter 4). Finally, the "thingness" of my shoes makes prosecuting theft much more straightforward: at least theoretically I can establish a causal chain between my original possession of the shoes and the thief's current possession of them. This chain is completed by the brute facts of identity: that pair of shoes is identical to mine, and is now in his possession. Since the raw material of IP is not physical, the idea of "chain of custody" makes almost no sense.

Analogies that stress the uniquely valuable character of works can go grotesquely awry. Consider Defoe's argument that his books are the "brats of his brain" (see chapter 4). He compares unauthorized copying to kidnapping children and selling them as slaves. But as Mark Rose points out, "if literary pirates are unchristian child-stealers, what are men who sell their children for profit?"[8] Defoe's analogy (that authorship is like paternity) clashes with his desire to defend an author's economic rights. Patterson and Lindberg argue that copyright is not ownership of a work, but ownership of a right: the exclusive right to make copies or derivative works based on a particular work.[9]

Seen in this light, the analogy to real property is quite limited. The fact that the author owns a "copy right" is logically independent of whether or not she has any other property rights. To claim that the analogies given above establish greater rights is simply to beg the question.

Arguments from Labor

The earliest justifications for IP rights were based on the author's labor. The Italian Giacopo Acontio made the following argument to the court of Elizabeth I in 1559: "nothing is more honest than that those who by searching have found out things useful to the public should have some fruit of their rights and labors, as meanwhile they abandon all other means of gain, are at much expense in experiments, and often sustain much loss."[10]

Acontio can be read as claiming that authors and inventors deserved to be compensated for their labor. However, this is far short of claiming that authors and inventors deserve to own the fruits of their labor. To posit such an ownership right would be to create a sort of "Marxism of the elite." Other workers aren't assumed to gain equity in what they make: why should authors?

Even the argument that authors deserve compensation to make up for giving up "all other means of gain" seems questionable if we make it a matter of economics. The arch-capitalist could certainly respond "I never asked him to quit his day job to write that symphony. He deserves my thanks for his effort, but I certainly don't owe him more than that. After all, he's presenting himself as an entrepreneur. Would I deserve something if I spent years inventing something that nobody wanted? The market rewards meeting a need: there is no "A for effort.""

More sophisticated arguments from labor are possible. In the following section we will examine two contemporary attempts to ground IP rights in labor.

Labor and Intellectual Property: Hughes and Yen

In "The Philosophy of Intellectual Property," Justin Hughes argues that appropriation subject to Lockean limitations is actually easier to understand in the context of IP than in the physical world. Moreover, "Locke's labor theory . . . can be used to justify intellectual property without many of the problems that attend its application to physical property."[11]

Hughes build his theory around the existence of an intellectual commons that provides the raw material for authorship:

> The "field" of all possible ideas prior to the formation of property rights is more similar to Locke's common than is the unclaimed wilderness. Locke's common had enough goods of similar quality that one person's extraction from it did not prevent the next person from extracting something of the same quality and quantity. The common did not need to be infinite; it only needed to be practically inexhaustible. With physical goods, the inexhaustibility condition requires a huge supply. With ideas, the inexhaustibility condition is easily satis-

fied; each idea can be used by an unlimited number of individuals.One person's use of some ideas (prior to intellectual property schemes) cannot deplete the common in any sense. Indeed, the field of ideas seems to expand with use.[12]

The inexhaustibility and abstraction of the intellectual commons is tempered by the requirement that ideas be physically realized before they have any real value. This process of realization is the key point where labor becomes relevant to the intellectual commons. Hughes argues that labor is the key to understanding the legal distinction between idea and expression:

> I suggest that the idea/expression dichotomy and the idea/execution dichotomy are the same. At a minimum, the force behind the latter dichotomy—the concern for labor—significantly contributes to explaining the idea/expression division. The courts' ad hoc approach in this area suggests that copyrightability may be based as much on what we feel are people's deserts as on what we feel are society's informational needs. It has been said that the idea/expression issue is uniquely well-suited for juries. I suggest that this is so not because juries care about a doctrine that ameliorates copyright and first amendment tensions and not because they know what idea-making is, but rather because jurors sense what labor is.[13]

Though Hughes puts great emphasis on labor, his theory is not (only) a labor theory of value. The author's labor deserves compensation because it creates social value: "This position holds that when labor produces something of value to others—something beyond what morality requires the laborer to produce—then the laborer deserves some benefit for it."[14] Hughes sees a hierarchy of value in the intellectual commons. This value becomes social utility by the process of expression, which involves labor. Thus "Protection of expression and not of ideas can be understood as protection for that part of the idea-making process that we are most confident involves labor."[15]

Though social value provides the ultimate rationale for awarding property rights, labor can provide the instrumental criteria for making IP decisions. For example, Hughes argues that derivative works do not deserve IP protection because its author "has not added much value (or much labor) beyond [original] idea X."[16]

Alfred Yen's paper "Restoring the Natural Law: Copyright as Labor and Possession" represents an attempt to make the implicit natural law undercurrent of copyright law come to the surface and do explanatory work.[17] Unlike Hughes, Yen is relatively hesitant to engage in extensive theory building: his primary focus is a close historical analysis of judicial decision making about copyright, followed by a critique of purely instrumental theories of copyright. The natural law principles he introduces make labor and "possession" (occupation) the natural basis for copyright. His natural law account is also unusual in that it focuses more or less equally on author's and user's rights.

Yen sees property claims in works as involving both the labor theory of value and the unconfinable nature of ideas. Yen argues that "The logic of natural law concepts like res communes and *ferae naturae* was that recognition of certain forms of property ran the risk of being an exercise in futility. It made no sense to make the air one person's permanent property because air could not be possessed. Similarly, when early American courts held that copyrights were inherently beyond the common law, they were essentially saying that the intangible nature of copyright made its possession inherently difficult."[18] Property rights in a commons must be tempered by a recognition of the fact that it is in the nature of many common things (e.g., the ocean, the idea of a hero's quest) to be difficult or impossible to enclose.

Yen argues that labor does create a natural claim to property rights. At the same time, he argues against the assumption that laboring automatically creates property rights themselves:

> The justification of property as the consequence of a person's labor implies that an author's labor of creation supports copyright. *As a general matter, a person rightfully claims ownership in her works to the extent that her labor resulted in their existence.* The property exists regardless of any need to economic stimulation of creative activity. However, it must be remembered that not all labor results in property. *To the extent that the author creates things which are not capable of possession under the law, the natural law prohibits the creation of a property interest.*[19]

Yen's account appeals both to an author's natural right, and the fact that some works by nature cannot be possessed. This helps Yen avoid the most common worry about introducing natural law into an IP theory, namely "the belief that natural law implies an unprincipled expansion of author's rights that will run amok over the public interest in free access to works."[20]

Hughes and Yen both make a strong case for taking labor as a central concept of IP theory. But they do not provide much comfort for author-centered theories, since both argue that the nature of ideas imposes strong constraints on the kind of property rights that labor can justify. Thus we find ourselves left at essentially the same place the analogical argument left us: whatever plausibility these arguments give property rights is undercut again and again by the "unphysical" nature of the intellectual world.

The next two arguments attempt to sidestep the nature of IP by focusing on particular features of the creative process.

The *Sine Qua Non* Argument

It could be argued that an author deserves to own his work because it wouldn't exist without him. Although many agents contribute to the publication of a work, only one represents a *sine qua non* for its existence. The claim that the author is a *sine qua non* seems to evaporate with the most casual scrutiny. Are we saying that no one else could have created the work? If we read "could" in a strong (modal) sense, the claim is patently false. If we read it in a weaker sense (something like "under the prevailing circumstances no one else could have created this work") it still seems questionable.[21]

We also need to ask whether being a *sine qua non* is transitive. If I am the *sine qua non* of my work, and my parents are the *sine qua non* of me, do my parents deserve to own my work? If not, what justifies drawing the line at the author? One thinks of an actress accepting an award, who spends an hour thanking everyone who has ever done anything for her. Can we ever say that none of these people provided a *sine qua non* for the actress' work?

The *Ex Nihilo* Argument

The *ex nihilo* argument goes as follows:

> Works belong to authors because authors create them "out of thin air." Since the creation of a work in no way depends on the property of others, it must be the property of the author.

This argument must face at least two serious problems. Despite the rhetoric of some artists, the first premise is not always true. Phone books and dictionaries are not created out of thin air, yet arguably deserve IP protection. Even when an author does create *ex nihilo,* the second premise only establishes that the work doesn't belong to anyone else, a fact that does not entail that the author owns the work. It would be equally consistent to argue that the *ex nihilo* creation of a work proves that it belongs to no one. Establishing the author's claim requires support either by an argument from labor or on the basis of some *sine qua non* argument.

The "Best of the Available People" Argument

A final argument for author-centered IP rights may appear to be no more than a desperate ploy:

> The utility of IP rights are beyond dispute. Property has to belong to somebody. Of the available parties, authors are the most plausible candidates for at least two reasons: (a) the author has a unique causal relation to the work and (b) the author deserves compensation for the labor that he has invested in the work. Granting ownership rights to authors also has positive consequences, by reducing the monopoly power of publishers and encouraging the efforts of new authors.

Despite the modesty of its premises (or perhaps because of them), I find this the most plausible argument for author-centered IP. Claim (a) is a restricted version of the *sine qua non* argument, and claim (b) is a restricted version of the argument from labor. These claims at least give plausibility to author's rights, though they aren't enough by themselves to establish them.

To my mind it is the consequentialist argument that is most decisive. The history of publishing prior to 1710 could be seen as an experiment in granting IP rights to publishers. The results were neither just nor economically efficient. Granting authorial control of works makes it much more likely that authors will be economically rewarded for their efforts, and it creates a countervailing force to the power of publishers.

The largest question still needs to be faced: does this argument justify recognizing the author's property rights as full liberal ownership?

Critiques of the Author-centered Theory

Work-for-hire

The 1909 Copyright act introduced a new concept into IP property rights: the concept of "work for hire."[22] The work for hire doctrine states that whenever a work is produced by an employee, the copyright is assigned to the employer. As Patterson and Lindberg point out, the doctrine has striking implications for the romantic theory of authorship:

- It effectively repudiates the claim that authors deserve ownership of their works because of the nature of authorship. Employee authors are presumably just as creative and original as freelance authors: yet they have no chance of owning their work.
- It undermines the idea that works are in some sense extensions of the author's personality. According to the doctrine, authors *prima facie* have no greater rights with respect to their intellectual works than any other employee has to the fruits of their labor.
- It severs the conceptual link between possession of copyright and the life of an author. Works for hire have a term corresponding to life of the author, yet the author has no legal connection to the work.

Authors have always been able to alienate their work: as we saw earlier, the original copyright system was built on the assumption that publishers are the natural locus of property rights. Work-for-hire is more radical since it breaks the link between author and property rights before a work is even created. Work-for-hire effectively erases the distinction between authorship and other forms of labor. This makes it impossible to argue for an essential connection between property rights and the activities of authors.

Practical Critiques

An author-centered theory may tend to deny some deserving claimants of IP rights because they aren't enough like the romantic author. The shamans who show bioprospectors medicinal plants are not "authors" and thus do not deserve compensation for their role in the development of new drugs. Another example is the case of *Moore v. Regents of the University of California.*[23]

Mr. John Moore had his spleen removed as part of treatment for cancer. Unknown to him, his physicians isolated a commercially valuable cell line from his spleen, which was patented by the University of California. When Mr. Moore learned of the commercial use of his cells, he sued for a share of the proceeds. The court ruled against him for two reasons. First, the defendants argued that the physicians were the true "creators" of his cell line, and that Mr. Moore was simply a form of raw material. This claim seems outrageous unless the test of intellectual property ownership is some form of creative transformation. Second, the judge accepted an argument made by the defendants that recognizing ownership of body parts would tend to undermine the development of medical knowledge by burdening researchers with the need to resolve ownership issues with every single subject.

An exclusive focus on the claims of authors encourages predatory behavior by "authors" who use the machinery of IP regimes to claim property rights. One example would be "cybersquatting." Since the right to domain names on the Internet are sold on a first-come-first-serve basis, it is possible for speculators to claim the names of products and corporations and then demand payment for releasing them to their appropriate owners. Perhaps the most egregious example of such grasping is the story of two doctors who independently developed a surgical technique, which only one wanted to patent.[24] The doctor who chose to forgo a patent found himself sued by by the other for using "his" procedure. He was easy to find: he had described the technique in the same issue of an opthomology journal as the patented technique.

Institutional Critiques

Purely author-centered theories of IP cannot accurately reflect the history of authorship, or even the history of intellectual property. As we saw in chapter 1, the first forms of copyright did not directly involve authors at all. Author-centered theories see protecting the economic rights of authors as the rationale for the entire IP system, but cannot explain the fact that authorship existed for millennia before anyone thought it necessary to develop IP laws. Something is missing here.

That something is the role publishing plays in the development of IP laws. It could be argued that despite all the talk about authors, IP laws do a much more effective job protecting publishers than they do protecting authors. A rationale for IP rights based on the romantic model of authorship also sits uneasily with the full alienation of IP rights when an author sells her work to a publisher, or the work-for-hire doctrine, which automatically strips an employee of any rights as an author and hands them to their employer. A cynic could argue that author-centered IP is a sham promoted by publishers to conceal their economic power behind a facade of moralism.

Strict IP Rights as Self-Defeating

According to the romantic model of authorship, an author's claim to his work turns on the originality of that work. Consider the difficulties of such a claim.[25]

Suppose that someone sues me for copyright infringement. If we make originality the touchstone of ownership, then they have to prove that they were original and that I wasn't. Either claim will be impossible to empirically verify in the vast majority of cases, without somehow exhaustively reconstructing the creative process for both of us. Any contact with my opponent's work, however tangential, would count against my claim of originality. At the same time, my accuser must somehow exhaustively account for all of the content in his work and show that it isn't derived from anyone else's IP. Without such absurd accountings, there would be no way to distinguish between conscious copying on my part, some kind of unconscious copying, or simply having a very similar idea for a work.

A purely author-centered theory is ultimately self-defeating. The rationale of an AC theory is to protect the rights of authors to profit by their works; but such profit is impossible unless a market for works exists and authors are legally allowed to create new works. It is important to remember that every author begins as a user, collecting material from an existing public domain to combine into a new work. Broad protection for IP rights will ultimately make it harder and harder for an author to publish a work without investing as much in litigation as in creating the work.

It might be objected that this is an argument against granting authors ownership of all aspects of their work, rather than an argument against an author-centered theory per se. This objection misses the mark: the essence of an author-centered theory is not that authors deserve ownership of every part of their work, but that the property rights of authors alone are the only important right a theory IP ethics needs to consider. The ultimately self-defeating character of strict property rights shows that author's property rights must be naturally constrained in order for the system of authorship to exist at all. But if authors must forbear ownership in some situations, then someone other than an author must have a right to protection against the author's claim.[26]

And if someone else has such a claim right, then author's rights alone are not adequate to ground all IP rights.

It is also important to remember that the interests of authors and publishers generally run parallel, even if they are not identical. The relationship is analogous to the relationship between the members of a union and the management of a company: each seeks to gain at the expense of the other, but each needs the other to survive. At this point it is impossible for authors to "fire" publishers or vice versa. Again, authors must forbear some property claims so that the system of authorship can survive.

This argument could also be directed against publisher-centered or user-centered theories of IP. The essence of the argument is that each member of the triad of authors, publishers, and users needs the other two, and that the interests of each group will never be identical to the interests of the other groups. We should ponder the status of this claim. It is not purely logical: it seems to presuppose facts about how market economies work and about how each party can be compensated. It seems logically possible that in some economic system authors, publishers and users could be distinct groups and yet somehow have identical interests. At the same time, if we accept as a matter of fact that the three groups have distinct interests, it seems to follow that no monolithic theory can adequately justify an IP system.

Notes

1. Michel Foucault, "What is an Author?" in *The Foucault Reader*, edited by Paul Rabinow (New York:Pantheon 1969).
2. Michel Foucault, "What is an Author?" 108.
3. Michel Foucault, "What is an Author?" 109.
4. Michel Foucault, "What is an Author?" 109.
5. Michel Foucault, "What is an Author?" 112.
6. We should point out that Foucault himself emphatically rejects the kind of theory described above: "these aspects of an individual which we designate as making him an author are only a projection, in more or less psychologizing terms, of the operations that we force texts to undergo, the connections that we make, the traits that we establish as pertinent, the continuities we recognize, or the exclusions we practice." Michel Foucault, *"What is an Author?"* 110.
7. That is, property rights in land ("real" estate).
8. Rose, *Authors and Owners*, 39.
9. Patterson and Lindberg, *Copyright: A Law of User Rights*, 12–13.
10. Price, *English Patents of Monopoly*, 7.
11. Justin Hughes, "The Philosophy of Intellectual Property," *Georgetown Law Journal* 77 (1988): 297.
12. Hughes, "The Philosophy of Intellectual Property," 316.

13. Hughes, "The Philosophy of Intellectual Property," 314.
14. Hughes, "The Philosophy of Intellectual Property," 305.
15. Hughes, "The Philosophy of Intellectual Property," 314.
16. Hughes, "The Philosophy of Intellectual Property," 318.
17. Alfred C. Yen, "Restoring the Natural Law: Copyright as Labor and Possession," *Ohio State Law Journal* 51 (1990): 517–559.
18. Yen, "Restoring the Natural Law," 551.
19. Yen, "Restoring the Natural Law," 534. Emphasis added.
20. Yen, "Restoring the Natural Law," 546.
21. This is especially clear in patent law, where companies can spend decades coming up with virtually identical inventions and litigating about them. Warshofsky's *Patent Wars*, 18–28, describes the endless competition between Procter & Gamble and Kimberley–Clarke over the design of disposable diapers.
22. Patterson and Lindberg, *Copyright: A Law of User Rights*, 85–89.
23. Boyle, *Shamans, Software, and Spleens*, 21–24.
24. Seth Shulman, *Owning the Future* (Boston: Houghton Mifflin, 1999), 38-43.
25. Litman, "The Public Domain," 1000–1003.
26. Though the argument at the beginning of the section is couched only in terms of authors, its point is that authorship presupposes "non–authorial" activities like access to the work of others. If these activities are legitimate for authors then it seems plausible to argue that *ceteris paribus* they are legitimate for anyone.

11

Ethical Issues of Patent Law: Equity and the Intellectual Commons

In this chapter we will be looking at the intellectual commons through the lens of patent law. One of the appeals of looking at patents is an opportunity to directly evaluate an IP system using the tools we developed for understanding common property regimes. Since patent regimes are based on precedents, they face issues like the conquistador problem whenever a new type of invention becomes patentable.

Patents are in many respects, the paradigmatic form of IP. Patents existed before copyrights and trademarks, and at least some of the distinctive features of the latter are derived from the details of patent law. Most of the differences between patents and the other forms of IP are in the direction of simplicity. Unlike copyright and trademark, patents can't be claimed: they must be won on a case-by-case basis in a process that requires the inventor to prove that his invention is useful, novel, and non-obvious. Patent claims can be opposed during the filing or invalidated by legal action after the patent is granted. The term of a patent is unconnected to the lifetime of the inventor, and the length of that term has hardly changed in the last two hundred years.

Patents are procedurally more complex than copyrights. However, at a conceptual level they are significantly simpler than copyrights. Copyright theory can be difficult to analyze because of the way that the author metaphor overshadows virtually everything else. It is tremendously difficult to get past the author in order to see anything else. Strangely, the inventor has never carried the cultural freight that authors have. They have been pushed aside by the long-standing Greek prejudice that saw manual labor as degrading. Except for a few cultural icons like Thomas Edison and Alexander Graham Bell, inventors are largely absent in historical narratives. The prejudices at work can be clearly seen in Catherine Macaulay's 1774 pamphlet *A Modest Plea for the Property of Copyright.* As Mark Rose describes it,

> she mightily objected to [his] classifying authors with inventors. Just as writers

had been unrealistically elevated—"with the intention of depriving authors of
the honest, dear-bought reward of their literary labors, they have been raised a
little higher instead of lower than the angels"— had been "levelled with the inventors of a very inferior order." The author was engaged in the improvement
of the human mind, whereas the inventor was concerned with the production of
luxuries or at any rate "conveniences, which are not absolutely necessary to the
ease of common life." Also there was a great difference in the way the products
of inventors and authors were received by the public. "Every common capacity
can find out the use of a machine; but it is a length of time before the value of a
literary publication is discovered and acknowledged by the vulgar."

The long-standing western prejudice against handwork was augmented by the
ideology of the author. Unlike writing or visual art, inventions reflect little or
nothing of the inventor's personality. They are almost totally constrained by the
functional rules of their domain of application. This is why the standard of originality is so much higher with patents than with copyrights. The genius of an inventor is his ability to solve a technical problem in a "non-obvious" way, to
come up with a new idea in the face of the constraints created by utility.

Attempts to explore copyright through concepts of natural rights find a clear
picture of author's rights and a blurred, almost imperceptible image of user
rights. The patent system almost reverses this bias: patents provide a tremendously clear picture of social value, combined with a blurred image of the inventor and his personality.

The scope of patentable material has undergone the same kind of broadening
we have already seen with copyright. In the last twenty years, patent protection
has been extended to organisms, genes, software, plant varieties, and surgical
procedures. There have even been proposals to allow the patenting of athletic
maneuvers.[1]

The most dramatic changes are in the increasingly abstract character of
patentable things, and the increasing ease with which previously existing things
can be turned into "inventions."

We will begin the chapter by taking a detailed look at biotechnology patents.
Expanding the scope of patents to organisms, genes, and sequences of DNA has
been tremendously controversial, and the debate has many implications for the
existence of an intellectual commons. We will look at several specific objections
to biotechnology patents, and then look at one commons-based argument: the
claim that genetic material is part of "the common heritage of humanity."[2]

In the second half of the chapter we will examine patents in light of the intellectual commons theory. Patent law exposes a different set of ethical issues than
does copyright law because patents cover useful things. Consequences of
changes in patent law can literally be matters of life and death in a way that
copyright decisions never are. And since the development of technology is inex-
tricably entangled with the progress of science, there is a constant struggle to decide what part of scientific discoveries can and should be owned. We will give

particular attention to three areas: the conquistador problem, anticommons problems, and the constraints on IP rights created by the rule of necessity.

The Human Genome and the Intellectual Commons

In 1972 Ananda Chakrabarty, a microbiologist employed by General Electric, used recombinant DNA technology to create a bacterium capable of breaking down crude oil. When the Patent Office rejected his application, he sued and begin a legal journey that led to the Supreme Court's 1980 *Diamond v. Chakrabarty* decision which extended patents to living organisms. The patent application had been rejected on the grounds that an organism was unpatentable as a "product of nature." This decision was based on 1948 Supreme Court decision that stated "patents cannot issue for the discovery of the phenomena of nature. The qualities of these bacteria, like the heat of the sun, electricity, or the qualities of metals, are part of the storehouse of knowledge of all men. They are manifestations of laws of nature, free to all men and reserved exclusively to none."[3] But by 1980 the Court had changed its mind. In a 5 to 4 decision, the justices ruled that since Chakrabarty's bacteria were not found in nature, they were "novel" enough to count as a new invention. In a famous statement, the Court gutted the 1948 decision by stating that "anything under the sun made by man" could be patented.

Chakrabarty involved patenting an entire organism: the next step was the development of arguments for patenting genes. At first sight, this seems significantly less likely: while Chakrabarty was in a position to argue that his combination of a naturally occurring organism with a gene from another organism was novel, it seems much harder to make a similar claim for genes themselves. People find genes: at this point, no one knows how to design one.

To get these under the patent tent, we need the further convention that a natural "arrangement of matter" can be patented if it is possible to physically isolate it from its natural context. This is a long-standing convention in patent law, and seems plausible for substances like penicillin. Fleming isolated penicillin, and in so doing he created something "artificial" that was novel and useful. The courts have used the same analogy with genes: "The researchers studying genes take a piece of DNA from its original source on a chromosome. They put it into a vector—a special segment of DNA arranged so that the inserted DNA can express a protein. The researchers then argue that they have then "isolated" the gene. Since genes are not present in the human body in precisely that form, they are viewed as man-made and potentially patentable."

In the world of patents it is only utility that tells the difference between invention and discovery. When the Supreme Court announced that anything under the sun made by man could be patented, they were in effect defining the ontology to be presumed in patent law. According to this ontology, what makes a thing, as it were, "patent considerable," is that it is "made by man": the only important distinction is between the natural and the artificial. As the Chakrabarty decision showed, the standard for "making" something can be as low as combining two natural things. The only relevant criterion for artificiality is whether or not that exact kind of thing already existed in nature.

Criticism of gene patents has often been couched in poetic, highly abstract language:

> To many people, the human genome is the sacred birthright of humanity, a priceless heritage that should not be desecrated by an ugly stampede to claim exclusive rights to vast collections of genes in the form of ESTs. "It makes a mockery of what most people feel is the right way to do the genome project," declared Stanford Nobel Laureate Paul Berg. . . . The French research minister, Hubert Curien, argued that "a patent should not be granted for something that is part of our universal heritage."[4]

This language seems to be saying that the genome should be part of an intellectual commons.

As we work with the following arguments against gene patents, we need to keep in mind the difference between the genome and individual genes. It might make sense to allow ownership of some members of a class while forbidding ownership of the entire class. If our goal is to promote progress, we must especially avoid granting IP rights to entire functional classes of genes.

Are Genes Discovered or Invented?

The first argument we will consider against gene patents goes like this: the genome is a pre-existing feature of the natural world. It seems an abuse of language to take credit for "inventing" something that already exists and whose existence in no way depends on human activity, "creative" or not. As we have seen, patent law accepts the principle that isolating a natural substance creates something that did not exist previously in nature and therefore can be patented. However, the argument does capture the unease felt by some members of the genomics community about patenting genes themselves.

The creation of biotechnology patents is easier to understand once we grasp how the terms "invention," "discovery," and "novel" are used in patent law. Ulrich Schatz offers the following clarification of inventions and discoveries:

> For the purposes of patent law, however, a discovery is a teaching which we have learned from observation and interpretation of nature, but which you cannot make or use for industrial purposes. But as soon as you put that abstract knowledge or teaching to actual use, so that a new man-made product becomes available, you have made an invention. Finding out that genes exist at all was a discovery. The identification of the sequence of the gene and the protein for which it encodes is also a discovery because it is still in the human body and you cannot make that protein, for instance human interferon or human insulin,

from a thing which is in the body of a living entity. But isolating it and putting it into a host cell gives it to you as a tool for the industrial production. What you have then is a new factory and that is an invention. . . . For patent law, which is about business and competition, the esoteric considerations whether a thing belongs to the sphere of fundamental science or natural science of the sphere of applied science, is of no meaning. A thing which you can make and which can be used in trade and industry is a possible object of a patent. We are dealing with business law, with patent law, and not with scientific theories.[5]

How Can a Gene Be Novel?

There a variety of ways to argue that genes cannot possibly be novel. For instance: some company now owns the human insulin gene. Does this mean that I infringe on their patent every time my pancreas uses the gene to make insulin for me? If it does, what do they expect me to do about it? In fact, the patents for genes only cover the isolated version of the gene, not the gene in its natural setting. This allows me to go on synthesizing insulin. But I would not be allowed to isolate my own insulin gene and use it to manufacture insulin in a laboratory.

Another version would go like this: there is no way to consider the insulin gene "novel," since it has been a functioning part of millions of people throughout history. While that is certainly true, we need to ask ourselves whether or not that means that the gene is a possession of its bearers. While I have an insulin gene and a unique genome, I don't know how it works, I wouldn't be able to directly use it for anything, and I wouldn't even recognize it if I saw it. I wouldn't be able to tell mine from anyone else's either (other than the fact that it is buried somewhere in my cells). My body is full of unfathomably complicated machinery that I take care of and that takes care of me. But it seems an abuse of language to call me the "inventor" or "maker" of that machinery. My right to my insulin gene seems rather more like the right of occupation: there it was, I needed one, nobody else can use it while I do, end of story. It seems as absurd for me to claim ownership of it on any other basis as for some genetic engineer to do so.

Patent law also has its own definition of "novelty." As R. Stephen Crespi puts it,

> Another argument against patenting genes is that, because of its pre-existence, a gene cannot fulfill the patent law test for novelty. But this test is framed in terms of availability to the public. Thus it focuses only upon what is already in the public domain through public disclosure or use prior to the filing of the patent application. Genes do not easily fit into this scheme. To be made available to the public the gene must first be isolated, preferably characterized as to its nucleotide sequence, and cloned. . . . Genes are therefore a special case of the broad class of naturally occurring substances which in appropriate circumstances can be patented.[6]

Genes and the "Inventive Step"

It could be argued that Craig Ventner has no more right than anyone else to own a gene. Billions of people already "own" copies of the gene themselves, donated by their parents and their parents before them. They would seem to have as much right as he does to claim that they are the authors of the gene. Thus we find ourselves returning to John Moore and his spleen.[7]

The case is especially significant in this context because it was the first to examine the questions of whether or not Mr. Moore owned either the cells used by his doctors or the genetic information encoded in them. The court's decision turned on both consequentialist and conceptual arguments. The court's unintentionally ironic appeal to problems property rights could create for medical research are discussed elsewhere in this chapter. Here we will sort through the court's conceptual arguments for denying Mr. Moore a property right in his genes. James Boyle argues that its real motivation was the belief that it was inappropriate to treat Mr. Moore as the author of his cells. He cites as evidence the following argument from the decision:

> Finally, the subject matter of the Regent's patent—the patented cell line and the products derived from it—cannot be Moore's property. This is because the patented cell line is both factually and legally distinct from the cells taken from Moore's body. Federal law permits the patenting of organisms that represent the product of "human ingenuity," but not naturally occurring organisms. Human cell lines are patentable because "long term adaptation and growth of human tissues and cells in culture is difficult—often considered an art" and the probability of success is low. It is this inventive effort that patent law rewards, not the discovery of naturally occurring raw materials.[8]

In effect, the appeals court said that Mr. Moore doesn't own the final product because he didn't engage in the creative labor that was necessary to create it. And it would also appear that he couldn't have owned his own cells because he didn't engage in any "inventive effort" to create them. They are simply "naturally occurring raw material." Moreover, he had "abandoned" them by consenting to have them removed from his body.[9] As Boyle puts it, "Mr. Moore is the author of his destiny, but not of his spleen."[10]

This line of argument assumes both that the romantic theory of authorship is correct and that genes or cells can only be owned as IP. Whatever the merits of the romantic theory of authorship, it seems that a much simpler case could be made against Mr. Moore's doctors. If someone steals the materials used to create a new invention, it would seem that they are guilty of theft whether or not they are guilty of patent infringement. Of course, this line of argument would neither say that the doctor's patent was invalid or that Mr. Moore was entitled to a percentage of the royalties.

Perhaps this situation somehow falls under the category of innocent use. It could certainly be argued that Mr. Moore was not harmed by the removal of his spleen, since that removal was medically necessary. Thus the use of his spleen cells did not in any way interfere with his own use of his spleen or do him any other harm (since he was not already selling his cells). On this line of argument, the doctors "read" Mr. Moore and used the information they found in him as they would read a book. Even if Mr. Moore is the author, they have done him no more harm than they would have done by reading any other book.[11]

Genes as Unauthored Ideas: People and Ideas as Raw Materials

If a path across someone's property becomes a major thoroughfare then he can lose the right to erect fences and stop the traffic. As Carol Rose describes it, "the several prescriptive doctrines for roadways, taken together, could act as a double-edged sword against the landowner. If the owner did nothing to halt the public's use, his passivity could be regarded as 'dedicating' the roadway to the public. If on the other hand he attempted to halt that use but failed, he could lose his rights under a theory of the public's 'adverse use'."[12]

When the courts act to protect the property rights of IP owners, they do it in the name of economic efficiency and also in terms of a higher public good. But courts can also make the same arguments to deny the property claims of others. Again, Moore's spleen offers a good example.

The court offered two kinds of arguments against Moore's claims to ownership of his cells and genes. Here we will examine the argument that granting Moore's claim would have negative consequences on research: "At the present, human cell lines are routinely copied and distributed to other researchers for experimental purposes, usually free of charge. This exchange of scientific materials which still is relatively free and efficient, will surely be compromised if each cell sample becomes the potential subject matter of a lawsuit."[13] Though the court made this argument against Mr. Moore, the judges were unwilling to turn it against the defendants. Later in the decision they exactly reverse the above argument: "the theory of liability that Moore urges us to endorse threatens to destroy the economic incentive to conduct important medical research."[14]

How could the court endorse both arguments? Boyle argues that "commodification of information can always be portrayed either as a time-consuming and unjust impediment to, or a necessary prerequisite for, the free flow of information." He traces this duality to the fact that economics treats the free flow of information both as a prerequisite for economic activity and as a commodity that can be treated as private property.

> The analytical structure of microeconomics includes "perfect information"— meaning free, complete, instantaneous, and universally available—as one of the defining features of the perfect market. At the same time, both the perfect and the actual market structure of contemporary society depend on information be-

ing a commodity—that is to say being costly, partial, and deliberately restricted in its availability. Our concern with market efficiency pushes us toward information flows that are costless, general, and fast. Our concern with incentives for the producers of information pushes in exactly the opposite direction—toward temporary monopolies that delay the release of information, limit its availability, and raise its price.[15]

If the goal of a particular decision is to promote particular forms of socially useful activity, then it might very well make sense to appeal to both arguments at the same time. In effect, the court is saying that Moore's property rights would inhibit medical research while the University of California's property rights would promote it.

This case is not about a matter of eminent domain: the court is trying to figure out if competing property claims even exist in this situation. Nor is the case primarily about whether or not cells or genes per se can be regarded as property, since the researcher's property rights are taken as relatively unproblematic. The central issue is whether or not a new kind of property right should be introduced into the legal universe: namely, the right to claim parts of one's own body as private property. If Moore's cells are not his property, then they are part of the commons that can be appropriated by the researchers to use as raw materials. If they are his property, then others must get permission for use and compensate him. The court rejects introducing what they perceive to be a new property right because it would interfere with the socially valuable activity of medical research. Others might feel that their decision opens the door to the commodification of human flesh itself.

It is not surprising that one group might feel their property rights threatened by the rights claims of others. Is it inconsistent or hypocritical? Only if we assume that everything should be private property. To the extent that the court's argument rings true, it is an argument for the necessity of an unowned source of valuable things: a commons. The researchers don't want Moore (or anyone else) to "fence off" his genes and thereby prohibit their use by anyone who can do something of social utility with them. There are two kinds of arguments for leaving his cells in the commons: (a) it would have bad consequences (the case discussed here) and (b) there is a morally significant difference between Moore's "natural" cells and the "artificial" cell line that the researchers created.

Genes as the Common Heritage of Mankind

Some researchers (and some governments) consider the genome to be part of the "common heritage of mankind" (CHM). In *The Concept of the Common Heritage of Mankind in International Law*, Kemal Baslar traces the development and uses of the common heritage doctrine.

The common heritage doctrine was first proposed by Maltese ambassador to the U.N. Arvid Pardo in 1967, as part of a debate on mining the deep sea bed. The deep sea bed is outside national borders and thus could be seen as part of an open-access commons. Pardo objected to simply allowing the industrialized nations to consume a resource that could plausibly be said to belong to everyone. Malaysian prime minister Dr. Mahatir Mohamad made the same point a bit more bluntly in 1982:

> The days when the rich nations of the world can take for themselves whatever territory and resources that they have access to are over. . . . Henceforth all the unclaimed wealth of this earth must be regarded as the common heritage of all nations on this planet.[16]

Baslar points out that there is an ambiguity in the claim that a resource "belongs to everyone." Pardo's original approach was to argue that all nations of the world deserve a share of any benefits that are derived from common heritage resources. Another approach is to argue that all nations of the world deserve access to the CHM itself and a voice in managing it.

Baslar argues that Pardo's original interpretation impeded the development of the CHM doctrine by identifying it socialist demands for wealth distribution:

> while Pardo was arguing that the Maltese proposal aimed to replace the principle of the freedom of the high seas with the principle of the common heritage of mankind in order to preserve the greater part of ocean space as a commons accessible to the international community, he tried to explain the common heritage of mankind by saying, at the initial stage of the UN Seabed committee, that it was a *"socialist concept."* To him, the common heritage of mankind, first and foremost, meant primarily the right to an equal share. This was such a feeble argument that the common heritage of mankind was challenged even by the then Socialist countries. One delegate from a Socialist country, for example, said at the first stage of the 1970 Declaration on the Deep Seabed that "obtaining profit without working for it is against socialism. It is just like an absentee-landlord theory."[17]

Pardo's argument went something like this: the deep seabed belongs to all nations equally. Therefore anyone mining the seabed is using "my" property and owes me compensation (and presumably must also get my permission and do whatever respect for my property rights would require). As stated, this argument faces a host of problems:

- Is the common property of mankind a joint tenancy? If so, why should the "silent partners" expect to share the profits without also sharing the costs? Pardo might respond that the basis of their claim to an equal share is based only on their right of tenure alone, but this is simply an endorsement of the principle of free riding.

- Pardo's claim that all countries have full and identical property rights in the seabed replaces the *res nullius* commons of the high seas with a full-blown anticommons where any holdout can stop any of the other partners from doing anything. This threatens to scuttle the whole point of insisting on property rights in the first place.
- By phrasing the argument in terms of an ideologically loaded, one-sided demand for compensation, Pardo is bringing conflict into the heart of the theory. The real and subtle concerns associated with sharing a common world get lost in cold-war confrontation.

The CHM model has been successfully applied to a number of entities, including the deep seabed, Antarctica, and outer space. A CHM regime combines the following features:[18]

1. The area cannot be annexed to national territory.
2. An international authority should manage all use of the areas and [access to] their resources will be shared equitably.
3. Any benefits arising from exploitation of the areas and their resources will be shared equitably.
4. The areas and resources are only to be used in peaceful ways.
5. The areas and their resources are to be protected and preserved for the benefit of present and future generations.

Seen through the eyes of commons theory, the primary function of a CHM regime is to guarantee that use of CHM resources are directly or indirectly consistent with the Lockean proviso. This can be accomplished in two ways. First, those who would consume CHM resources must directly compensate latecomers and others for the loss of "as much and as good" (benefit sharing). The other alternative would be to treat the CHM resource as a strict commons, allowing only renewable use (or in some cases allowing no use at all), and mandating a positive obligation to preserve it for future generations. Thus a CHM regime can be interpreted as a kind of commons.

This interpretation must not blind us to the distinctive features of the CHM approach. In particular, the CHM approach focuses heavily on reconciling state sovereignty and the equitable exploitation of resources outside states. The CHM model was originally developed to keep certain places from being appropriated by nations. In Grotian terms, places like Antarctica are transformed from *res nullius* to *res communis* by a CHM regime. The CHM model seems essentially tied to geography, making its application to other domains problematic.

The CHM understanding of benefit-sharing is also problematic. As Baslar has pointed out, a concern with equity opens the door for replacing common use rights with rights to compensation. While this may address some short-term issues of distributive justice, it undermines the common nature of the system as a whole. Being paid off by the first to arrive is in effect selling them your common use rights. As the critics of the enclosure movement stressed, one-time compensation does not adequately compensate a commoner for the loss of all potential future use of the commons.

Problems with Applying Common Heritage Concepts To the Human Genome

In her article "Sovereignty and Sharing," Bartha Maria Knoppers describes her proposal to incorporate common heritage claims into the UNESCO *Universal Declaration on the Human Genome and Human Rights*.[19] She was only partially successful: the final draft of the declaration included the statement that "The human genome underlies the fundamental unity of all members of the human family, as well as the recognition of their inherent dignity and diversity. In a symbolic sense, it is the heritage of humanity."

According to Knoppers, there were four types of objections to claiming that the genome is part of the common heritage:

> While the International Bioethics Committee embraced the common heritage of humanity concept, certain government representatives mandated to study and approve the Committee's draft Declaration understood the common heritage concept as mandating possible appropriation by international conglomerates, and thus, like in the Rio debate, a risk to state sovereignty. Free market advocates disliked the community aspect and others fearful of State sovereignty wanted to protect the human genome at the individual level. Finally the French translation of heritage as "patrimony" also created difficulties since it would be seen as having an economic meaning. Hence, the adoption of the watered-down expression "symbolic of the heritage of humanity."[20]

Knopper's opponents had wildly divergent political agendas, but all of their complaints seem to be derived from the fact the genome is already inside national (and even personal) boundaries. Those on the left feared that a CHM regime would keep governments from regulating biotechnology, and might even allow someone to claim some kind of property right in another person's genes.[21] Those on the right feared that a CHM regime would prevent appropriation of genes and exclusive ownership of biotechnology. The failure of the genome to be outside the system of nation-states means that a CHM regime would inevitably conflict with existing national and personal rights. No one will sign on for such conflicts without a convincing demonstration that the benefits of a CHM regime would outweigh its costs. Unless some variation on CHM can be developed, it does not seem to be a useful tool for criticizing biotechnology patents.

Patents and the Intellectual Commons

The Conquistador Problem in Patent Law

Since the patent system is based on the use of precedents to limit patent claims, it stands to reason that when a new art is opened up to patent the first claimants can make very broad claims indeed. If those claims are allowed, all successive inventors will be constrained by the need to respect the "foundation" patents (at least until the patents expire).

The "Harvard Oncomouse " patent is a good example of this problem.[22] "The description that was provided by the applicant, Harvard University, only contains information about a method to produce onco-mice, whereas the claims are directed to *all non-human mammalian onco-animals.*"[23] Though the Harvard researchers had only succeeded in placing human oncogenes into a strain of mice and had done no experiments on other animals, the title of their patent application was (in part) "Method for producing a transgenic non-human animal having an increased probability of developing neoplasms." In effect, Harvard was attempting to patent purely hypothetical applications of their methods of producing oncomice.

This situation represents a form of the conquistador problem we discussed in chapter 5. Initial appropriation is unconstrained in a completely new domain, and this creates the potential for a claimant to appropriate so much that it unjustly deprives other commoners and undermines the commons itself.

The intuition behind these concerns involves a kind of basic fairness. It seems unfair for one person to profit from the work of another and claim it as their own, especially when the sole base of that property claim is a willingness to assert exclusive rights through the legal system. The patenting of surgical procedures offers a particularly egregious example:

> The downside of patenting medical procedures drew attention when Dr. Samuel Pallin patented a method for performing cataract surgery and in 1995 sued Dr. Jack Singer for using the technique without paying a royalty . . . in the Pallin litigation Jack Singer ultimately won because he could show that Pallin had patented a technique already in use. Singer himself had performed the operation a month before the patent claim. But the verdict in his favor could not make up for the vast sums of money he spent in litigating the case—nor in the fact that many patients were denied this beneficial treatment while litigation took place.[24]

Why was Pallin granted a patent in the first place? It now seems obvious that "his" technique was already in use (at least by Singer!). As long as the the process of evaluating the novelty of a patent claim is primarily based on searching earlier patent claims, the potential for "land grabs" is enormous. The problem is especially serious in disciplines like surgery and computer programming, where skills are transmitted by oral tradition and apprenticeship. Without a written record, it is impossible for the patent office to prove that someone didn't invent a particular technique or device.

Broad patents can also be used as a kind of economic land mine. The current generation of "patent trolls" scour patent records for broad patents that are not being enforced. They then buy these patents and sue any possible infringer. Many companies will pay them to go away. A company called TechSearch is a typical example:

> In the last three years, TechSearch has made millions of dollars–primarily from a patent it acquired on a method of transmitting data between computers. Close to 100 companies—including UAL Corp., Sears Roebuck and Co., and Hyatt Corp.—have opted to pay TechSearch to license the technology rather than take the fight to court. . . . TechSearch embodies the criticism leveled most against patent enforcement specialists: It is not the inventor who sought the patent, it produces nothing, it sells nothing. It simply makes money by exploiting broad patents that have never really been enforced. It lives, primarily, to sue.[25]

In an article in the Harvard Business Review, Kevin Rivette and David Kline exhort their readers to "deploy their patents not just as legal instruments but also as powerful financial assets and competitive weapons."[26] They advocate a variety of tactics including "bracketing": if a competitor fails to patent all the required features of an invention, patent them yourself and use the patents to prevent him from producing his product.[27] They also approvingly describe how the company Avery Dennison used patents to shut down a competing product line: "we went to Dow and basically told them that they couldn't manufacture that film anymore. They had to shut down their team, dismantle it, and withdraw it from the market. And that's exactly what Dow did. . . . we were able to stop Dow in the market and have it basically all to ourselves."[28]

In one sense such bare-knuckle tactics reflect the fact that IP grants are (and are intended to be) monopolies. The problem with tactics like bracketing is that the company seeking the bracketing patent is exploiting the patent system itself, rather than being rewarded for the invention of something useful. It is also not difficult to see that such tactics will almost automatically create an anticommons, since each inventor has an incentive to protect himself as much as possible from others by seizing all the patent rights he can.

In an extensive study of patent scope, Robert Merges and Richard Nelson conclude that granting broad patents has a negative impact on the process of invention itself, though the impact is industry— and technology—specific.[29] Broad patents have a particularly negative effect on emerging technologies (ironically the technologies patent regimes are designed to encourage).

Necessity and the Commons

In *On the Laws of War and Peace* Grotius argues that individual property rights can be overridden in cases of dire necessity: "if a man under stress of dire necessity takes from the property of another what is necessary to preserve his own life, he does not commit a theft."[30] As he points out, the same right had been recognized by Aquinas among others.[31] Grotius takes great pain to point out this is an active right on the part of the person in need, not a duty owed by the property holder:

> in direst need the primitive right of the user revives, as if community of ownership had remained, since in respect to all human laws—the law of ownership included—supreme necessity seems to have been excepted. . . . The reason which lies back of this principle is not, as some allege, that the owner of a thing is bound by the rule of love to give to him who lacks: it is, rather, that all things seem to have been distributed to individual owners with a benign reservation in favor of the primitive right. For if those who made the initial distribution had been asked what they thought about this matter they would have given the same answer we do. "Necessity," says Seneca the father, "the great resource of human weakness, breaks every law" meaning of course human law, or law constituted after the fashion of human law.[32]

He goes on to hedge this principle with various limitations: the necessity must be unavoidable,[33] the right does not apply when the possessor is in equally dire straits,[34] and that there is an obligation to replace the things taken whenever possible.[35]

The argument from necessity seems pretty clear when failing to get food or water creates a risk of imminent death. But does it have a more attenuated form? Robert Nozick's discussion of the "Historical Shadow" of the Lockean proviso considers this question at length.[36]

At first glance, Nozick does not seem to have much sympathy for such a possibility: "But a right to life is not a right to whatever one needs to live; other people might have rights over these other things. . . . At most, a right to life would be a right to have or strive for whatever one needs to live, providing that having it would not violate anyone else's rights."[37] However, this is not the end of the story. The Lockean proviso that appropriation must leave "as much and as good" for others means that even if no one has a claim on me for the things necessary for life, I have an obligation not to leave the world in such a state that they would be unable to pursue appropriation of the things necessary for life on their own.[38]

The same constraint applies to existing property: "If the proviso excludes someone's appropriating all the drinkable water in the world, it also excludes him purchasing it all."[39]

Nozick justifies this constraint by appealing to the "historical shadow" of the Lockean Proviso. This shadow is functionally similar to Grotius' "benign reservation" but not equivalent to it, since Grotius (unlike Nozick) believes that necessity can override property rights.

Necessity, Market Failure, and Government Research

Pharmaceutical companies argue that if they are forced by compulsory licensing to sell drugs cheaply in the developing world, they will have no incentive to develop treatments for diseases of the poor. This makes sense in terms of the incentive model, but ignores an important reality: they aren't developing treatments for diseases of the poor anyway.[40] Public goods problems are not the only potential source of market failure. It is not economically rational to develop medicines for people so poor they can't buy them. In the developed world, market failure can be due to a limited market: it makes no sense to develop expensive drugs to treat "orphan" diseases that only affect a few hundred people every year.

Ironically, the IP system itself can cause market failure. Solomon Snyder has argued that American drug companies were slow to introduce lithium treatment for bipolar disease into the American market because lithium (as a natural element) is unpatentable.[41] As the patent for Prozac came close to expiration, Lilly introduced "r-isomer," "once-a-week," and "pediatric" Prozac.[42] The only proven advantage for any of these compounds is that they allow Lilly to seek a new set of patents. Eflornithine, the best available drug for treating sleeping sickness, was withdrawn from the market in 1995 for financial reasons: it was recently reintroduced as a treatment for preventing the growth of facial hair on women. The NGO Doctors without Borders recently succeeded in shaming the manufacturer into resuming production for sleeping sickness treatment.[43]

My intention in raising these examples is not to castigate pharmaceutical companies. The point is that private pharmaceutical companies, like all private companies, exist to make money. Quite naturally their research efforts will follow profit before need. Thus defenders of IP cannot argue that protecting or increasing property rights is the solution to the problems of distributive justice associated with pharmaceutical research. Nor, I think, can they argue that limiting their IP rights will reduce their incentives to treat the diseases of the poor: the market has already done that.

Governments can attempt to ameliorate pharmaceutical market failure in a variety of ways. The least intrusive method would be to artificially subsidize the development of "unmarketable" drugs through guaranteed purchases, tax breaks, and the extension of stronger IP rights to drug companies (this is the approach taken by the Orphan Drug Act). The most radical method would be to either legally compel their production through compulsory licensing, or for the government itself to manufacture the drugs (ignoring existing IP rights in the process). Both of these approaches attempt to solve the problem from the "demand" side: it is also possible to work on the problem from the supply side.

A Commons-based Solution

The central economic problem of pharmaceuticals is the need for enormous capital investment in the development process: it can cost hundreds of millions of dollars to develop a new drug. Governments can (and do!) address that problem at least partially through public research and development. "Upstream" basic research that is freely available gives "downstream" pharmaceutical companies a way to reduce the cost of producing drugs.[44]

For this approach to work economically, the publicly generated resources must be either free or greatly discounted. The resources in question (scientific and technical knowledge) are different from public infrastructure like roads, since it can be appropriated by individuals to use in the creation of private property. And if they are free or greatly discounted, justice demands they be available to any potential developer. Government funded basic research thus can help overcome market failure by contributing to the intellectual commons.

Several conditions must be met in order for public basic research to solve market failure problems. First, the public research must focus most intensively on areas most severely affected by market failure issues. Other areas of basic research (like particle physics or cosmology) will require less pragmatic justifications. Second, we must have some way of requiring a social benefit for the social cost of basic research.

The Threat of Creeping Privatization

Congress passed a series of laws in 1980 (the Bayh-Dole Act and the Stevenson-Wydler Act) and 1986 (the Federal Technology Transfer Act) that allow federally-funded universities, government agencies, non-profit organizations and federal employees to patent their inventions. The laws also provide tax incentives for private corporations to fund research and allow the creation of "cooperative research and development agreements" (public–private partnerships). The distinction between "pure" or "basic" research, funded by the government, and "applied" or "downstream" research pursued by private companies for commercial advantage is being erased. The net result is a process of privatization that steadily shrinks the domain of public knowledge.

The growing privatization of research has two kinds of adverse consequences. It undermines the norms of information sharing that are a basic part of scientific research, and creates an anticommons that impedes both commercial and academic research. The decline of public research also exacerbates the problems of market failure: it further reduces incentives to search for treatments for diseases of the poor.

Conclusion

A theory of common property rights needs to reconcile individual appropriation with the preservation of the commons itself. One of the simplest methods is to distinguish between the commons and its fruits: a commoner can appropriate apples, but not the apple tree. We could even extend this to positive duties: apple eaters have a duty to plant new apple trees.

Imagine a commons where apples are unknown. A traveler moves to the commons and brings a bag of apple seeds with him. The commoners are simple folk who don't understand the connection between seeds and apple trees. He plants an apple orchard and gives the other commoners access to apples. Is he obligated to share the seeds and his knowledge of horticulture?

The traveler could certainly argue that he has no such obligation. He has not taken anything away from the other commoners: in fact, they are better off than they were before. In some narrow sense this seems like a reasonable argument. But in another sense, it is very unreasonable indeed. The commoners are no longer in their pre-apple situation. There is a method of extracting value from the commons that only the traveler has access to. If he were to leave and all the trees were to die, the commoners would no longer have as much and as good.

The traveler deserves compensation for his contribution to the commons. But by planting his trees, he has changed the commons itself, and it will not still be common if he withholds his secret forever. We also need to recognize that his contribution does not consist in creating apple seeds *ex nihilo*: his labor and knowledge are only part of the story.

Notes

1. Consider the following whimsical example from Sport Illustrated: "The Bob Cousy BEHIND THE BACK PASS. The inventor, a fully accredited basketball professional, seeks to protect a form of offensive chicanery in which he manually maneuvers a basketball behind his back, sometime accompanying the maneuvers with deceptive movements of his head and/or eyes, and rapidly discharges said ball so as to reassign possession of it to a teammate." Quoted in Andrews and Nelkin, *The Body Bazaar*, 205, note 106.

2. Kemal Baslar, *The Concept of the Common Heritage of Mankind in International Law* (Norwell, MA: Martinus Nijhoff, 1998).

3. *Funk Brothers Seed Co. v. Kalo Inoculant Co.* 333 U.S. 127 (1948).

4. Kevin Davies, *Cracking the Genome: Inside the Race to Unlock Human DNA.* (New York: Free Press, 2001), 63.

5. Ulrich Schatz, "Patents and Morality," in Sigrid Sterckx , ed., *Biotechnology, Patents, and Morality* (Brookfield, VT: Ashgate, 2000), 226.

6. R. Stephen Crespi, "The Case for Patenting Biotechnological Inventions," in Sigrid Sterckx, ed., *Biotechnology, Patents, and Morality* (Brookfield VT: Ashgate, 2000), 286.

7. Boyle, *Shamans, Software and Spleens*, 97–107.

8. Boyle, *Shamans, Software and Spleens*, 106.

9. This in spite of the fact that leaving them in his body would have been fatal, and

the fact that he was never informed of their intended use!

10. Boyle, *Shamans, Software and Spleens*, 107.

11. Though we must exercise at least a little caution here. Had Mr. Moore "published" his body by taking it to the doctor, or by walking around in public?

12. Carol Rose, "The Comedy of the Commons" in Carol Rose, *Property and Persuasion* (Boulder, CO: Westview, 1994), 114–115.

13. Boyle, *Shamans, Software and Spleens*, 101.

14. Boyle, *Shamans, Software and Spleens*, 101.

15. Boyle, *Shamans, Software and Spleens*, 101.

16. Baslar, *Concept of Common Heritage*, 35.

17. Baslar, *Concept of Common Heritage*, 31–32.

18. Baslar, *Concept of Common Heritage*, 35.

19. Bartha Maria Knoppers. "Sovereignty and Sharing." In Caulfield and Williams–Jones, eds., *The Commercialization of Genetic Research: Ethical, Legal, and Policy Issues* (Dodrecht: Kluwer, 1999).

20. Knoppers, "Sovereignty and Sharing," 3.

21. "paging Mr. Moore. Your spleen is on the phone, and it's laughing."

22. Sigrid Sterckx, "European Patent Law and Biotechnological Inventions," in Sterckx, ed., *Biotechnology, Patents, and Morality,* (Brookfield, VT: Ashgate, 2000), 22–25.

23. Sterckx, "European Patent Law and Biotechnological Inventions," 23. Emphasis in original.

24. Andrews and Nelkin, *The Body Bazaar*, 53–54.

25. Brenda Sandburg, "Battling the Patent Trolls," *The Recorder*, 2.

26. Kevin Rivette and David Kline, "Discovering New Value in Intellectual Property," *Harvard Business Review* 78, no. 1 (Jan. 2000): 54–66.

27. Rivette and Kline, "Discovering New Value," 58.

28. Rivette and Kline, "Discovering New Value," 63.

29. Robert Merges and Richard Nelson, "On the Complex Economics of Patent Scope" *Columbia Law Review* 90 (1990), no. 4.

30. Grotius, DJBP, 2.2.6.4.

31. Thomas Aquinas, *Summa Theologica*, II-ii, 66.7.

32. Grotius, DJBP, 2.2.6.4.

33. Grotius, DJBP, 2.2.7.

34. Grotius, DJBP, 2.2.8.

35. Grotius, DBJP, 2.2.9. This last condition is a bit puzzling, since Grotius assumes that the person in need has a right to take what they need. If they have a right, why do they have to compensate?

36. Nozick, *Anarchy, State, and Utopia*, 178–182.

37. Nozick, *Anarchy, State, and Utopia*, 179.

38. The weak principle he uses is "A process normally giving rise to a permanent bequeathable property right in a previously unowned thing will not do so if the position of others no longer at liberty to use the thing is thereby worsened." Nozick, *Anarchy, State, and Utopia*, 178.

39. Nozick, *Anarchy, State, and Utopia*, 179.

40. Doctors Without Borders, 2001. "What is the Campaign for Access to Essential medicines?" www.doctorswithoutborders.org/advocacy/access/crisis.html.

41. "Even at this point, the drug was slow catching on, especially in the United States. Drug industry economics may have been a factor. The major antischizophrenic and antidepressant drugs, introduced to psychiatry in the mid–1950s, were all patented

chemical entities . . . Lithium, being a well–known metal ion, was not patentable. Thus, it is hardly surprising that the major drug companies were reluctant to spend the many millions of dollars required for toxicity studies and clinical trials before such a product could be brought to market. Not until the mid–1960s was lithium marketed commercially in the United States and abroad, finally benefiting hundreds of thousands of patients afflicted with mania." Solomon Snyder, *Drugs and the Mind* (New York: Scientific American Library, 1986), 116.

42. Lilly has also introduced the "new" drug Serafem (Prozac) for the treatment of pre–menstrual syndrome.

43. Donald G. McNeil, "Profits on Cosmetic Save a Cure for Sleeping Sickness," *New York Times*, February 9, 2001.

44. See Heller, "Can Patents Deter Innovation?" for a good discussion of upstream and downstream research.

12

Conclusions and Critique

This final chapter will be devoted to examining various criticisms of the theory of intellectual commons, looking at some alternative approaches suggested by some of the moral problems examined earlier, and closing with some final reflections on taking control of the evolution of IP. The theory described in the preceding chapters is considerably messier than a purely economic and author-centered theory. Here we will confront the army of objections that the complexity of the theory invites.

Does Time Heal All Wounds in IP Ethics?

The fact that all IP grants eventually expire might seem to be the real saving grace of morally questionable IP regimes. After all, every monopoly right (just and unjust) will eventually be canceled by the passage of time. Though patenting plants may be outrageous, the patents will expire in twenty years. Everything created by authors will become part of the commons eventually. So where is the problem?

This argument won't fly, for several reasons:

1. Even if the street belongs to everybody (as thus also to me), I don't have the right to set up a roadblock and extort "tolls" from anyone who wants to get past. The fact that I quit doing this after April 7 doesn't make what I'm doing right before April 7. Time limits are irrelevant to the fact that I'm preventing others from using something that is as much theirs as it is mine.
2. Limitation of term can be subverted by periodically extending the period of protection until it become effectively permanent ("eternity by installment").
3. Expansion of the subject matter of IP creates a permanent class of

things that can be taken out of the commons, and this global effect is not time-limited even if each individual grant is time-limited.
4. There is a clear and universal movement in the evolution of IP rights: the movement toward more and longer.[1]
5. The single-mindedness of this development is not what one would expect if the process were being driven by an attempt to balance individual rights and social utility. The most plausible interpretation is that IP rights are slowly evolving into property rights *simpliciter*. I have argued elsewhere that this movement needs to be opposed, since it would undermine the original purpose of IP grants, namely the advancement of knowledge. My point here is that if this process continues, limitation of term will survive as a purely formal "fig leaf" concealing the reality of an empty intellectual commons.

The claim that time heals all wounds in IP is equivalent to the claim that limitation of term alone is sufficient to create a just IP regime. This claim is false for both theoretical reasons (point 1above) and practical reasons (points 2–4 above). Limitation of term is only an instrument, a means. A viable IP regime requires some specification of what the end of the system is, and what countervailing interest can justify limiting the property rights of authors. This unifying vision is best supplied by the concept of an intellectual commons.

The Geographic Metaphor

Some of the most salient features of the IC theory flow from its use of geographic metaphors. This opens the way for several lines of criticism.

The Geographic Metaphor Is Useless

There is room to argue that an appeal to the idea of a commons is really just to recycle the distinction between the public and the private. We can follow the logic of such an objection by following James Boyle's analysis of the public-private distinction in *Shamans, Software and Spleens*.[2] Boyle characterizes the distinction between public and private as a distinction between the moral and political regimes in operation:

> The liberal state depends on the idea of equality. That, after all is one of the key differences between the liberal and the feudal idea of politics. Liberalism mandates an end to status distinctions in politics. There can be no restriction of the franchise to a particular social class, no weighing of the votes of the nobility. Thus we have equality, but only inside the public sphere. Citizens are equal, but only in their capacities as citizens, not as private individuals. Each is guar-

anteed an equal vote, but not equal influence. We draw a line around certain activities—voting, appearing in court, and so on—and guarantee equality within this realm. Outside of the line is the private sphere, the world of civil society. It is the private sphere that contains all the real differences between people— differences of wealth, power, education, birth and social rank. It is this process of conceptual division that allows us to use the language of egalitarianism to defend a society marked precisely by a highly stratified distribution of wealth or power.[3]

Boyle sees a problem with trying to address distributive justice issues by rearranging the boundaries of the public/private distinction. Any such attempt will give rise to perfectly symmetrical moral arguments for moving the boundary in either direction, neither of which is *a priori* superior to the other. If we assume that both current property holders and the currently dispossessed have basic rights, then any change in the boundary will violate somebody's rights. Without appeal to some justification from outside the distinction itself there is no way to resolve the conflict.

For this argument to work against the IC theory it is necessary to identify the commons with the liberal public realm. But there is no obvious reason to do this, since we have argued that the commons is a realm of "natural" rather than conventional rights. Boyle's point seems to be something like this: we can't assume that developing a theory of the commons will solve any specific moral problems, because the boundary between public and private is not "natural" in the sense of being determined by the geographic metaphor itself. The boundary is a consequence of additional constraints that are not part of the public/private distinction itself.

Disanalogies Between the Intellectual Commons and the Environment

There are some striking disanalogies between the environment and the intellectual commons. Perhaps the most significant one is that the environment has a robust, independent, and inescapable physical reality. No one in a finite and well-populated world can ignore it. IP theory is built totally on analyzing high-order social and political institutions that do not seem to have any reality independent of human activity.

Decisions about how to treat the environment also have an objective basis, because we understand how the environment works as a physical and biological system.[4] In contrast, even the friends of the intellectual commons agree that it has an abstract and underdetermined character (with the enemies not even agreeing that it exists).

There is no way to go to the intellectual commons the way one can go to the rainforest. In fact, the intellectual commons seems to be a convenient fiction:

nothing but a shorthand way of talking about who has certain property rights and who doesn't. Why not just talk about property rights then?

I contend that the idea of an intellectual commons is a useful fiction, because it focuses attention on the unique status created for works when their term of ownership is over. These works are now available to anyone for use, but cannot be owned by anyone. They are either non-property or some form of non-individual property. The fact that this unique status exists invites reflection on whether there are moral reasons for extending it to other kinds of works. Maybe some kinds of knowledge are so intrinsically valuable that they should be exempted or excluded from ownership. Maybe some forms of expression are so profoundly moving that as many people as possible should be exposed to them. Maybe the foundational knowledge of a new technology should be exempted from ownership and developed by groups or governments. In all these cases, we are talking about the possibility of applying a model of property that is significantly different from either full liberal ownership or collective ownership through the state. That model, whatever it is, is what we call the intellectual commons.

The rise of environmental law was driven by (sometimes controversial) changes in attitude toward the natural environment but implemented in (generally accepted) institutional changes. The real legacy of environmental law is not universal consensus on the intrinsic value of nature but universal acceptance of the principle that environmental costs cannot be silently externalized. The intellectual commons approach to IP will be successful if it creates acceptance of the principle that IP policy changes must be evaluated pluralistically, instead of silently externalizing the costs of granting IP owners more rights.

Disanalogies Between the Intellectual Commons and the State of Nature

The attempt to apply natural rights or natural law concepts to the intellectual commons is also open to question. Intellectual property isn't "natural" at all: it does not exist independently of the state, and is a relatively recent idea. How can it possibly help to talk about natural rights or natural law when discussing IP?

There are two points to make in answer to this objection. First, natural law theories are not required to reductively explain all social institutions in terms of human nature or the state of nature. Though such a direct approach might be possible, it is also possible to appeal to natural law as a regulative or constitutive principle. On this view, the purpose of a natural law theory is to answer questions about the moral foundations of a given social or political order. Do people have rights and obligations that are not dependent on a grant from the state? Are social and political institutions a free human invention, or are they constrained in various ways by the kind of things people happen to be? The answers to these questions don't have to dictate the specifics of a polity, but they may be necessary to justifying its legitimacy.

The dominant strain of natural rights theory is perhaps the most extreme possible expression of moral individualism. Nozick lays his cards on the table in the first sentence of *Anarchy, State, and Utopia*: "Individuals have rights, and there are things no person or group may do to them (without violating their rights). So strong and far-reaching are these rights that they raise the question of what, if anything, the state and its officials may do."[5] Yet even Nozick admits that there are property rights that cannot be allowed because they foreclose too many possibilities for others.[6]

Grotius, Locke and the Scottish moralists all made sociability as basic a feature of human nature as individualism. Grotius and Locke both posited common rights as part of the state of nature and argued that these common rights persisted under governments. And both also made the concept of a commons a central feature of their analysis of property. So some kind of Grotian/Lockean natural rights theory offers conceptual tools for building a theory in which there are more natural rights than there are individual rights or property rights.

Moral Obligations To Create and To Share

Are All Moral Obligations Associated with IP Supererogatory?
The cornerstone of the author-centered theory of IP is that sharing works is a superogatory, not obligatory, act. Authors deserve praise for creating and sharing works, and a prudent state rewards this generosity with economic rights. A pluralistic theory seems to turn this on its head: authors have an obligation to share their work, and are morally culpable if they fail to share on generous enough terms. We will consider whether or not such obligations are unfairly singling out authors in the next section. Here we will explore the consequences of considering authors to have obligations to share their work. Our analysis will be driven by a series of test cases. Consider the following scenarios:

1. A brilliant but thoroughly misanthropic scientist, working alone, discovers a cure for malaria. He destroys his notes and tells no one. When he dies, the secret dies with him.
2. A brilliant but thoroughly disorganized scientist, working alone, discovers a cure for malaria. On his way to a meeting to announce his results he is killed in an accident. His notes are incomprehensible, and the secret dies with him.
3. A brilliant but thoroughly greedy scientist, working alone, discovers a cure for malaria. Realizing that the vast majority of those with malaria are impoverished, he concludes that he can't make money on the cure. He tells no one.

4. An brilliant but thoroughly greedy scientist discovers a cure for malaria. He patents it but refuses to produce it, in hopes of extorting huge subsidies from governments.
5. A brilliant but thoroughly misanthropic scientist discovers a cure for malaria. He patents it but refuses to produce it because he hates people (especially ones with malaria).
6. The scientist in scenario (5) patents his cure and also many of the crucial steps leading to it (he builds a "patent wall"). He refuses to allow anyone to work any of his patents. His actions put his cure out of reach and seriously hamper the search for alternative treatments.

Which of these scenarios involve moral culpability? If authors have sole, despotic dominion over their works then none of them do. Yet my intuitions tell me that these scenarios are not morally the same. Even though the first five have the same consequences, there seems to be a clear moral difference between (2) and the others, since the bumbler had intended to tell others of his discovery.

The traditional approach to capturing these moral intuitions is to make a distinction between perfect and imperfect moral duties.[7] In the first five cases, the scientists are failing to perform their imperfect duty of benevolence by withholding the cure. It would be morally praiseworthy for them to disclose it, but they aren't morally blameworthy for failing to disclose it. (The sixth case seems a bit different, since the scientist is actively impeding the efforts of others to help and well as refusing to help.)

Is making this distinction enough to resolve the moral unease created by the scenarios above? It doesn't for me. There are two considerations not adequately addressed by treating disclosure of works as superogatory. The first is that the scenarios above describe what could be interpreted as a moral catastrophe.[8] The potential good that disclosure would accomplish is so great that non-disclosure doesn't just return the situation to a neutral baseline. It would seem to be possible to argue that on consequentialist grounds alone failure to disclose the cure is no different than driving a water truck up to people dying of thirst and then driving off without giving them any water.

While Nozick himself suggests that the truck driver is doing something wrong, he balks at drawing the same conclusion about the scientist:

> The fact that someone owns the total supply of something necessary for others to stay alive does not entail his (or anyone's) appropriation [of that resource] left some people (immediately or later) in a situation worse than the baseline one. A medical researcher who synthesizes a new substance that effectively treats a certain disease and who refuses to sell except on his terms does not worsen the situation of others by depriving them of whatever he has appropriated. The others easily can possess the same material he appropriated; the researcher's appropriation or purchase of chemicals didn't make the chemicals scarce in a way that would violate the Lockean proviso. Nor would someone else's purchasing the total supply of the synthesized substance from the medical researcher . . . This shows that the Lockean proviso is not an "end-state

principle"; it focuses on a particular way that appropriative actions affect others, and not on the structure of the situation that results.[9]

Nozick is making two points with this argument. The first is that his reading of the Lockean proviso is not consequentialist. What makes the truck driver's actions wrong is that his appropriation forecloses the possibility of appropriation of others and is thus a violation of their rights. The second point is that the scientist hasn't performed an act of appropriation, since the substance he created is not found in nature. Since it was not previously available for appropriation, he is not infringing anybody's natural rights by refusing to share it except on his own terms. Sufferers from the disease are free to pay for their own materials and research to duplicate his cure.[10] (Patents would complicate this picture, and Nozick follows this argument with some ideas about how the patent system should work, which we will consider later.) Works are never under the constraints of the Lockean proviso, though their material might be, since their value is imposed by creating something that did not exist before. The originating nature of invention exempts it from the rules of justice in acquisition and makes it perhaps the ultimate form of private property.[11]

A more general version of this argument would be something like this: I can only harm someone by preventing them from appropriating something from the commons without leaving "as much and as good." Since the medicine didn't exist at all before I created it, it was never part of the commons. If it isn't part of the commons, they have no *prima facie* claim to it. So there is no way that any action of mine can make their situation worse (though I could make it better). They are no worse off than if I hadn't been able to find the cure in the first place (or if I had never been born). Therefore sharing it is not obligatory.

However, the assimilation of inventions to private property seems a bit too hasty here. Couldn't it just as easily be argued that since the invention is an idea, and since sharing ideas automatically satisfies the Lockean proviso, I have no reason not to share it? After all, no matter how many times I let someone light their taper at mine, mine is still lit. The nonexclusive nature of ideas would seem to guarantee that no one can do me wrong by taking an idea of mine, since I still have as much and as good. Nozick here falls prey to the duality pointed out by Boyle above: his argument that those he didn't share with were not being harmed since they could duplicate his work independently can be turned around and used to ask why he should own the invention (since he has already admitted that others could produce it independently).

The Principle of Innocent Use Is Inconsistent with an Obligation to Contribute to the Intellectual Commons

It can be argued that the central role that innocent use plays in a theory of the intellectual commons means that it is impossible to argue for an obligation to contribute to the commons.

Consider the following argument:

1. If no one is harmed by my use of the intellectual commons, then my right to use it is unconditional.
2. Therefore, that use cannot be conditional on any reciprocal act on my part.
3. In particular, it cannot be conditional on some duty to contribute to the commons.
4. But the only possible ground for such an obligation is as a condition for use.
5. Therefore, there can be no obligation to contribute to the commons.

The argument here is that since it is impossible to harm anyone by using the materials of the intellectual commons, there is no reason to impose any conditions on that access; in particular, there is no reason to impose some duty to contribute to the commons. But if that follows, then contributing to the commons must be superogatory, and we are back where we started.

What are we to make of this argument? Premise (1) as stated seems to equivocate on the word "use." In the antecedent, we are assuming that the particular use in question is harmless, but in the consequent "use" seems to have grown into all possible uses. (1) can only be true as stated if all uses of the common are innocent (which they are not), or if all my uses are innocent. If the first, then the argument is unsound; if the second, then we seem to be saying something like all non-harmful uses of the IC are OK: reasonable, but not exactly earth-shaking. (1) also seems to be asserting a necessary connection between non-harming and unconditional rights. Yet innocent use seems neither necessary or sufficient for arguing for an unconditional right. It is not necessary, since it is possible to have an unconditional right to consume or destroy something. Treating the possibility of innocent use as entailing an unconditional right of access seems unreasonable.

The biggest problem for this argument is premise (4). It seems arbitrary and even perverse to argue that the only possible ground for an obligation to contribute to the IC is as a precondition for use. This seems arbitrary because there might be other good reasons for such an obligation (a variety of consequentialist ones spring to mind). It seems perverse because the non-depletable character of the intellectual commons means that it is not subject to free-rider problems: it is perhaps the only kind of common that isn't subject to overuse. There is no potential tragedy, since no act by other commoners can deprive an author of the use of his work. If the commons is a pasture, it makes sense to impose conditions on use, or even to restrict access to a group of users small enough not to exceed the carrying capacity.[12]

The IC Theory Picks on Authors

One line of argument against an IC theory is that it imposes unfair burdens on the producers of works. Joseph Addison painted the following sad picture in 1709: "he that has separated himself from the rest of Mankind, and studied the Wonders of the Creation, the Government of his Passions, and the Revolutions of the World, and has an Ambition to communicate the Effect of half his life spent in such noble Enquiries, has not Property in what he is willing to produce, but is exposed to Robbery and Want, with this melancholy and just Reflection, That he is the only Man who is not protected by his Country, at the same Time that he best deserves it."[13]

The claim is a familiar one. Authors are already making enormous personal sacrifices to create works in the first place: it adds insult to injury to impose unique obligations to share on them. A similar argument was made by the authors of the *NII White Paper* when they denounced the idea of "taxing copyright owners, apart from all others" to promote universal access to the Internet. Edward Samuels is offended by the clause in the Statute of Anne requiring that authors send copies of their books to the university libraries: "Quite a public works statute! If authors wanted the rights granted by Parliament, they had to donate books to all the major libraries." Samuel's disapproval seems to reflect the intuition that no one should have to pay for rights: rather, the burden is on others to recognize and honor them.

One must be careful not to push this argument too far. The first, and least sympathetic response, is Rousseau's response to a similar argument: "I didn't ask you to work, so I shouldn't have to pay you in the form of property rights for the work you did."[14] Society rewards people for creating value, not simply because they put a lot of time and effort into making something. A person who spends years squaring the circle doesn't deserve the rewards that are showered on a person who spends years (successfully) proving Fermat's last theorem. For the rest of this section we will focus on value as a criterion for desert, rather than labor *per se*.

The moral and legal principle at work here is the rule that I don't have to pay for a benefit I didn't ask for. Wendy Gordon explains it as follows:

> One of the supposed certainties of the common law is that persons need not pay for benefits they receive except when they have agreed in advance to make payment. The rule takes many forms. One of the most familiar is the doctrine that absent a contractual obligation, a person benefited by a volunteer ordinarily need not pay for what he has received. This rule supposedly both encourages economic efficiency and respects autonomy. To illustrate the baseline rule: If my neighbor drains his swamp and in the process also dries up the mosquito haven in my backyard, I am benefited. Nevertheless, the law will require me neither to shoulder part of the drainage costs, nor to hand over to my neighbor any portion of the increase in land value which his actions have given me.[15]

One must exercise a certain discipline when discussing subsidies and unfairness. One could argue that the fact that others profit from the actions of authors is unfair. But so is the market: except in the case of a monopoly, others decide the value of the fruits of my labor. I recently heard a congressman object to the federalization of airport security by saying that it would "create another group of federal workers to subsidize." I must disagree: a salary is not a subsidy, even for a federal employee. Changes in the legal and technological environment that affect the value of intellectual property are not necessarily injuries or benefits that someone is liable for. A really unsympathetic response to authors would draw a comparison between authors whose work ends up in Napster with steelworkers who lose their jobs because of foreign competition. Perhaps authors also need help retraining so they can go into another line of work.[16]

Reifying the Intellectual Commons

It could be argued that trying to make a strong analogy between environmental ethics and information ethics will only work if we reify the intellectual commons. The reason that environmental ethics is unique is that it articulates moral duties that are somehow owed to the environment itself. Nature has its own given structure and, in some fairly strong sense, its own needs that underpin environmental obligations. But to make similar kinds of assumptions about the intellectual commons seems to be opening the door to a bunch of truly strange obligations. Could there be an equivalent on the IC side of the principles of deep ecology? At the very least, it seems to lead to obligations to things like ideas and to the commons itself.

Are ideas sitting around in some platonic universe waiting to be expressed? Do we owe them that? If we fail to educate the young in their cultural heritage, are we doing an injury to the heritage itself, as well as to our young? We have some general idea of what the environment needs. What does the IC need? Doesn't its inexhaustible character put it beyond any sense in which we can hurt it? Should we embrace a principle of plenitude, and say that the IC needs us to make more and more things for it to hold?

The analogy between nature and the intellectual commons breaks down when we try to push it in the moral direction that deep ecology pushes environmentalism. There is a clear sense in which the Earth is independent of human activity, and there is also a clear sense in which the biosphere embodies a teleology that is independent of human purposes and goals. It was here a long time before we were. The intellectual commons doesn't have that kind of independent reality. It also seems to lack any kind of intrinsic telos that would justify treating it as an independent thing. The commons isn't a field—it's more like a warehouse. Its inventory is completely the result of human activity.

We don't need to reify the commons, and the fact that the commons is not independent of human activity in no way means that it isn't an important category. The descriptive character of the commons is secondary to its value: that hinges on its normative character. Developing a theory of the intellectual commons isn't doing geography (even if we use geographic metaphors). Rather, it is an attempt to systematically understand what things should be owned and which should be exempted from ownership. Adjusting the boundary between ownable and unownable is never simply a matter of surveying the hills and planes of the intellectual commons. "[T]he idea/expression dichotomy is fully dependent on a judge's characterization of what constitutes an 'idea.' By characterizing something as an 'idea,' a judge is essentially ruling that it is something that cannot be commodified for purposes of private ownership, but rather should be commonly shared. No certain rule can capture this inquiry."[17]

Reifying Works

It could be argued that the idea of an intellectual commons will lead inevitably to the bizarre step of reifiying works as something independent of specific physical things. It would open the door to idea that works themselves are somehow morally considerable. Do works have a right to life? Is the destruction of the last copy of a book the moral equivalent of allowing a species to go extinct? Is rejecting a manuscript like abortion? Are those who swap disks of abandoned programs protecting something like "infodiversity"?

If the commons has some kind of intrinsic value, that value must derive from the value of the "things" that constitute it (whether they are ideas or works). So it would appear that works have intrinsic value. But if they have intrinsic value, then their destruction or neglect is a moral wrong. Thus we find ourselves driven to assigning works some kind of independent moral considerability.

John Milton seems to be making just this kind of leap in *Areopagitica*. He reifies works in some truly startling statements:

> For books are not absolutely dead things, but do contain a potency of life in them to be as active as that soul was whose progeny they are: nay, they do preserve as in a vial the purest efficacy and extraction of the living intellect that bred them. I know they are as lively, and as vigorously productive, as those fabulous dragon's teeth; and being sown up and down, may chance to spring up armed men. And yet, on the other hand, unless wariness be used, as good almost kill a man as kill a good book. Who kills a man kills a reasonable creature, God's own image; but he who kills destroys a good book, kills reason itself, kills the image of God, as it were in the eye. Many a man lives a burden to the earth; but a good book is the precious life-blood of a master spirit, embalmed and treasured up on purpose to a life beyond life. 'Tis true, no age can restore a life, whereof perhaps there was no great loss; and revolutions of ages not not oft recover the loss of a rejected truth, for want of which whole nations fare the worse.[18]

It is true that in the next paragraph he backs off from this line of argument "lest I be condemned of introducing licence, while I oppose licensing." It is clear that

Milton didn't intend to be taken literally, but he clearly seemed to be arguing that works have some kind of intrinsic value that makes destroying them morally blameworthy.

What are we to make of this argument? It seems to me that there must indeed be some sense in which works, once created, take on a kind of life of their own. After all, artists can't have a moral right to preserve the integrity of a work unless there is something that constitutes such integrity. And since works somehow embody meanings that transcend the author's intentions, it does make sense to talk about them as "things." Perhaps the value we attribute to works can be parsed in terms of the needs and desires of users, though I am generally skeptical of such translation projects. In any case, the author and user independence of works does not require a commitment to some sort of platonic ontology. The loss of a work, like a death, is the loss of something valuable; but in this world such losses are often inevitable and unavoidable.

Intellectual Property as Solution and Problem

Those who embrace the vision of an intellectual commons are united in their belief that the existing owner-centered theory is both morally and practically defective. Some critics would argue for the dismantlement of the entire system.[19] Others expect (or at least hope) that the system will wither away of its own accord.[20] Still others see a need for large-scale changes in practice that leave the basic framework of IP intact.[21]

Which solution seems appropriate depends largely on the locus of the problem. Perhaps the benefits and costs of the IP system are maldistributed. Perhaps some who deserve rights have been left out. Perhaps we need to graft new institutional forms onto the existing corpus of IP law. The basic choice is between reforming the system and dismantling it—and our choice here is determined by how we answer three questions.

Have the Benefits of the Existing IP System Justified Its Costs?

We have alluded earlier to the maddeningly subjunctive nature of arguments about intellectual property. People on both sides of every question can make plausible arguments about the likely consequences of new policies, but short of actually enacting them, there is no way to find out who is right. The sharpest predictions are made by those with the narrowest theoretical focus, like Moglen's " econodwarves."

What Problems Do IP Systems Solve? What Problems Can They Solve?

It's hard to see how IP can make any positive contribution to distributive justice issues. The creation of a new "biorentier" class in developing countries is no more likely to promote distributive justice than oil elites have in the Middle East. Protecting indigenous art and culture through existing IP mechanisms risks making "aboriginal" just another brand. Those who want to protect cultural values shouldn't turn them into just another product. An assertion that traditional cultures are somehow the common heritage of humanity risks creating the same kind of conflict between local people and the world community that is emerging from the Rio protocol on biodiversity. IP is not a good way to solve issues of privacy (because we don't want privacy to be alienable in the same sense that IP rights are). Even in the matter of creating economic incentives, we must beware the claim that "more must be better."

An important issue that belongs somewhere: should the protection of a work be tied to how much it costs to produce? Should "big art" and "big science" get more protection that little art and little science?

What Problems Will *Limitations* on IP Solve?

If we want a scientific commons, there are two possible routes: (a) ban some types of patents or (b) explicitly codify and formalize and fund (or promote and nurture) such a system instead of simply remaining asleep at the wheel.

Allowing people to own their organs would at least force a quantum improvement in informed consent procedures. If we do this (maybe combined with statutory rules a la compulsory licensing), there should be no chilling effect on research.

The Future of the Intellectual Commons

The steady growth of IP rights cannot continue indefinitely. At some point, the utility of the system as a whole will cease to outweigh the dead-weight costs it imposes, and pressure for change will begin to build. The greatest risk to the system as a whole is the desire of publishers to go it alone and the obscene eagerness of some in the government to criminalize research and impose technologies of control on the public.

Even this will ultimately fail. The commons will always exist, and it will always be in someone's interest to keep it common. Works are not made from nothing. No one can predict what someone may figure out to do with one of their ideas. Like all that we make or do, it ultimately escapes us. There is comfort in this thought: at some point, the future has to take responsibility for using them and quit blaming us. When our names and our lives are forgotten, the ideas we expressed will live on, the common soil of humanity.

Notes

1. William Fisher, "The Growth of Intellectual Property: A history of the ownership of ideas in the United States," (September 1997), www.law.harvard.edu/Academic_Affairs /coursepages/tfisher/iphistory.html.

2. Boyle, *Shamans, Software and Spleens*, chapter 3, "The Public and Private Realms."

3. Boyle, *Shamans, Software and Spleens*, 25–26. Boyle credits this approach to Karl Marx (*On the Jewish Question*).

4. This is not to say that it is easy or obvious how to solve environmental problems, especially since our knowledge of things like climate change is still growing. Which thing to do may not be obvious, but the range of options is something that all competent participants could be expected to agree on eventually.

5. Nozick, *Anarchy, State, and Utopia*, ix.

6. Nozick, *Anarchy, State, and Utopia*, 55n. Nozick argues that my enemies shouldn't be able to trap me by purchasing all the land around mine.

7. This distinction is present in embryonic form in Grotius' distinction between perfect and imperfect rights, and is explicitly developed by Pufendorf. See Schneewind, *The Invention of Autonomy*, 133–134.

8. Nozick, *Anarchy, State, and Utopia*, 183–185.

9. Nozick, *Anarchy, State, and Utopia*, 181.

10. "There is no problem of homelessness: everybody just needs to buy a house."

11. Ayn Rand, "Patents and Copyrights," *Objectivist Newsletter* 3 (1964), no. 5.

12. The classic method of managing a "common property resource"; see Ostrom, *Governing the Commons*, chapter 1.

13. Rose, *Authors and Owners*, 36–37.

14. Jean–Jacques Rousseau, *A Discourse on the Origins of Inequality*, 354.

15. Wendy Gordon, "Touring the Certainties of Property and Restitution: A Journey to Copyright and Parody," 1.

16. In a cartoon by Ruben Bolling, a poet finally faces economic reality: "My dream of making millions through poetry is dust. Perhaps I should talk to my brother about that tire-sales job."

17. Gordon, "Touring the Certainties of Property," 22.

18. John Milton, *Areopagitica: A Speech for the Liberty of Unlicensed Printing*, Volume 29 of *Great Books of the Western World*, (Chicago: Encyclopedia Britannica, 1991), 384. Tacitus disagrees: "And so one is all the more inclined to laugh at the stupidity of men who suppose that the despotism of the present can actually efface the remembrances of the next generation. The persecution of genius fosters its influence: foreign tyrants, and all who have imitated their oppression, have merely procured infamy for themselves and glory for their victims." Annals, IV.35

19. Brian Martin, "Against Intellectual Property."

20. Eben Moglen, "Anarchism Triumphant."

21. This is how I read Lessig's *The Future of Information* and Boyle's "Belliago Declaration" in *Shamans, Software, and Spleens*.

Appendix A

Economics: Some Definitions

Common property resource: A common property resource is shared by a clearly delineated set of commoners, who mutually limit each other's consumption. This arrangement keeps any individual from completely externalizing the impact of their consumption, thus avoiding the tragedy of the commons.

Efficiency: An economic system is production-efficient if there is no way to increase economic output without reallocating resources in use. A production-efficient system neither wastes resources or allows them to go unused if their use would increase productivity. An economic system is distributionally efficient if it maintains a Pareto-optimal distribution of goods.

Externality and externalizing: North and Thomas define externalities as follows: "Private benefits or costs are the gains or losses to an individual participant in any economic transaction. Social costs or benefits are those affecting the whole society. A discrepancy between private and social benefits or costs means that some third party or parties, without their consent, will receive some of the benefits or incur some of the costs."[1] In the absence of common rules, individuals have an incentive to externalize as many of their costs as possible (e.g., overgraze pastures, dump pollutants, steal ideas rather than invent). This can lead to the tragedy of the commons, or to situations like the prisoner's dilemma, where it is impossible to maximize individual utility through the pursuit of self-interested actions.

Free-riding: A free-rider treats the property of another as if it were a public good. For example, a drug company that simply copies a successful drug gains the economic benefits of the drug without having to incur any of its costs. If free-riding is unrestricted, it becomes economically irrational to invest in research or the development of new products. This creates a kind of intellectual tragedy of the commons (see below).

Market failure: A situation in which market mechanisms are unable to maximize efficient production and distribution of resources. The two most significant examples are public goods (see above) and monopolies. A monopoly causes market failure because absence of competition largely frees a producer from the need to adjust prices to demand.

Market transaction: A market transaction takes place between a buyer and a seller, each of whom is attempting to maximize his own utility. If we take the manufacture and sale of some item as paradigmatic, the transaction can be modeled as a set of inputs and outputs. The producer has two types of inputs: those he must acquire through economic transactions (direct costs), and those he does not have to acquire through economic transactions (externalized costs). The producer and consumer also face costs associated with the transaction itself (transaction costs).

Monopoly and oligopoly: A monopoly exists when a good is only available from a single seller. In this situation, the lack of competition insulates the seller from market pressures that tend to control prices. Oligopoly is a situation where there is a small, closed group of sellers instead of a single seller. Even without direct collusion, all sellers can increase their income by simply raising their prices whenever a competitor raises theirs. Private property rights can be seen as an attenuated form of monopoly: the reason my holdings have *any* economic value is my ability to withhold them from the market. Private property doesn't create market failures because there are so many proprietors who have the same thing to sell. Intellectual property just is the creation of a market for works by investing some agent with state-enforced monopoly rights over the distribution of works.

Open access resource: An open access resource is a "maximally common" resource: it may be exploited by anyone without paying some owner. There are also no institutional limits on the amount consumed by each individual user. This leads inevitably to the tragedy of the commons (see below).

Pareto optimality: A distribution is Pareto-optimal if there is no way of increasing the welfare of some individuals without decreasing the welfare of others. Note that Pareto optimality is a far weaker measure of social welfare than the principle of utility. Pareto optimality is consistent with a low total utility (e.g., where most people are at a level of bare subsistence and a few are extremely wealthy) and a low average utility (since the initial distribution is taken for granted, which values the utility of initially favored individuals more than the utility of initially unfavored individuals).

Perverse incentives: Markets create incentives and disincentives for economic choices. The harmony of such incentives is the ultimate source of the legendary efficiency of markets. These incentives are the formal equivalent of selection pressures in evolutionary theory. One primary cause of extinction is over-specialization: *local* selection pressures drive a population to optimize in a particular direction, leaving them unable to adapt to significant *global* changes. Marx's most important contribution to economics is the encouragement to look at the *global (long-term)* consequences of responding to *local (short-term)* incentives. He argued that capitalism contained "contradictions," in the sense that pursuing a purely *laissez-faire* economic policy creates conditions like monopoly, which undermine the *laissez-faire* system itself. Sometimes following a locally optimal course of action will create conditions that will eventually lead to a situation where there are no locally optimal choices left. The tragedies of commons and anticommons are examples of such situations. We will use the term *perverse incentives* to describe the forces that drive an agent into choices that tend, in the long run, to undermine the system that made the initial choices possible. Perverse incentives are a logical, perhaps inevitable, consequence of dissonances between individual costs and social costs.

Problem of comparative utility: Some economists have challenged the use of the principle of utility in economics on two grounds. First, it is argued that since happiness is a mental state, there is no objective way to compare the utility of two individuals. Second, it is argued that the potentially redistributive implications of the principle of utility involve introducing "normative" baggage into the business of economics. The consensus solution to this problem has two parts: (a) move in a more behaviorist direction, replacing the concept of utility with that of preference satisfaction (since preferences are presumably manifest in behavior) and (b) use Pareto optimality as a global criterion instead of some global measure of utility.

Public goods: Public goods are goods that cannot be distributed to individuals and which individuals cannot buy a bigger share of by spending more. Mill's classic example of a public good is a lighthouse: it is visible to everyone whether they have paid a lighthouse fee or not, and there is no way for anyone to control more of the lighthouse's output by paying more.

Rent-seeking: Rent-seeking is the opposite of free-riding: a rent-seeker treats a public good as if it was private property. The classic English example is demanding a toll for the right of way:

> "Hold up!" said an elderly rabbit at the gap. "Sixpence for the privilege of passing by the private road!" He was bowled over in an instant by the contemptuous Mole, who trotted along the side of the hedge chaffing the other rabbits as they peeped hurriedly from their holes to see what the row was about.[2]

Rent-seeking appears in several forms in intellectual property. The most egregious example is patent extortion, where a claimant tries to be the first to file a vague patent that will give them the ability to threaten later inventors with litigation. The patent extortionist uses the machinery of the patent system to gain economic advantage without actually adding to the progress of the arts and sciences. Using technology to prevent legitimate "fair use" of IP is also a form of rent-seeking. Unrestrained rent-seeking increases transaction costs and can can lead to a situation known as the tragedy of the anticommons (see below).

Tragedy of the commons: A term coined by Garrett Hardin in his paper of the same name. In an open-access system, each user can retain the individual benefits of consumption while externalizing the costs. In a common pasture, the shepherd who keeps a bigger flock keeps the increased benefit, while "everybody else" absorbs part of the cost of overgrazing. If all the shepherds try to maximize their self-interest, they will rapidly destroy the common resource they all depend on. Hardin saw only two possible solutions: dividing the commons (which forces each user to internalize his resource costs) or the imposition of limits by some superior authority ("Leviathan"). Carol Rose and Elinor Ostrom have argued that common-property resource models offer a third alternative. Formally, the tragedy of the commons is a close relative of the N-way prisoner's dilemma.

Tragedy of the anticommons: Heller and Eisenberg define the tragedy of the anticommons as as situation in which "multiple owners each have a right to exclude others from a scarce resource and no one has an effective privilege of use."[3] This situation is a mirror of the tragedy of the commons: instead of a resource being used up, it is underutilized or unutilized. Examples given by Heller include the recent privatization of the soviet economy where according to Heller,

> transition governments often failed to endow any individual with a bundle of rights that represents full ownership. Instead, fragmented rights were distributed to various socialist-era stake-holders, including private or quasi-private enterprises, worker's collectives, privatization agencies and local, regional, and

federal governments. No one could set up shop without first collecting rights from each of the other owners.[4]

If open access makes free-riding economically rational, an anticommons makes rent-seeking economically rational. In the short term, this greatly increases transaction costs. In the long run, it creates a sort of "dog in the manger" climate, where no one is willing to give up any part of their control for fear of being out-performed by someone else. Many critics of current IP regimes fear that the expansion of IP rights will ultimately create an "intellectual anticommons" where new authors are so hampered by the rights of existing authors that the progress of science and industry will be slowed down.

Transaction cost: defined by North as "the costs of using the market to organize an economy."[5] North identifies three types of transaction costs: (a) search costs (the cost of bringing buyer and seller together), (b) negotiation costs (the cost of working out an agreement), and (c) enforcement costs (the cost of enforcing agreements that go beyond simple face-to-face exchanges). Efficient markets tend to reduce transaction costs. Enforcement costs are particularly important for intellectual property regimes. If the cost of enforcement exceeds the value of the goods exchanged, the transaction is not economically viable. This argument is the foundation of most economic arguments for state enforcement of IP rights as a public good.

Notes

1. Douglass North and Robert Thomas, *The Rise of the Western World: A New Economic History* (New York: Cambridge University Press, 1973), 2.

2. Kenneth Grahame, *The Wind in the Willows,* reprint of 1908 edition. (Farmington Hills MI: Charles Scribner, 1960) 2.

3. Michael Heller and Rebecca Eisenberg, "Can Patents Deter Innovation?: The Anticommons in Biomedical Research," *Science* 280 (1998), 698.

4. Heller, "Can Patents deter innovation?" 698.

5. North, *The Rise of the Western World,* 135–136.

Appendix B

Property, Ownership, and Rights: A Framework

We can distinguish between general and specific philosophical theories of property. The subject of a theory is the question "What moral principles justify any form of property?" A specific theory assumes the legitimacy and coherence of the concept of property, and addresses the moral justification of particular systems of property rights. We can construct a list of questions that any specific theory of intellectual property needs to address:

- What is the relationship between a theory of IP and theories of other forms of property? Can the different kinds of theory be developed independently, or must IP theory be reducible to some other understanding of property?
- Can we develop a theory of property that is coherent at the boundaries of ownership? That is, a theory that contains a defensible account of why some things are owned and other things are not, and how things can move across the boundary in a systematically meaningful way.

As we begin our analysis, we also need to be mindful of three cultural presuppositions in the western understanding of property.[1] Failing to question or even notice these presuppositions can make analyzing the concept of property significantly more difficult.

1. In the West, the default form of ownership is private property. There is an almost irresistible cultural tendency to view *all* forms of property through the lens of individual ownership, and to assume that deviations from private ownership are what a theory of property needs to explain.
2. In the West, property rights have what Charles Donahue refers to as "agglomerative tendencies." The basic understanding of ownership is in terms of a *single* individual rights-bearer, the owner.
3. Western legal systems have tended to make *possession* of something a central criterion for ownership. Who currently has something often

trumps any other consideration of how this state of affairs came about.

All three of these presuppositions can complicate developing a general theory of intellectual property and we will need to be mindful of them in what follows. For example, if ownership is possession, then IP can't be owned because there is no meaningful sense in which it can be "possessed." The agglomerative tendency is one source of the central tension of IP theory: the pull between a single owner and a vast world of users. Taking private property as paradigmatic also makes it very difficult to develop an positive theory of either common property or the un-owned.

Hohfield's Typology of Rights

We will approach property somewhat obliquely, by first analyzing rights and then the concept of ownership. One very useful framework is Wesley Newcomb Hohfeld's typology of rights as developed in his book *Fundamental Legal Conceptions*.[2] Hohfeld interprets rights as a relation between a right-holder, the object of the right, and other agents. He identifies four main types of rights: claim rights, liberties, powers, and immunities.

Claim Rights are rights that automatically entail the existence of moral duties. Becker's definition is typical, though he stresses enforcement rather than moral judgment:

> The existence of a right is the existence of a state of affairs in which one person (the right-holder) has a claim on an act or forbearance from another person (the duty-bearer) in the sense that, should the claim be exercised or in force, and the act or forbearance not be done, it would be justifiable, other things being equal, to use coercive measures to extract either the performance required or compensation in lieu of that performance.[3]

Any reasonable theory of moral behavior needs to recognize the existence of choices not constrained by moral imperatives. Hohfeld refers to such choices as **Liberties**. A liberty is a right that does not entail the existence of duties. More positively, a liberty is the power to make autonomous individual choices. I have a liberty, all other things being equal, to wear a white shirt or to step over cracks in the sidewalk. The concept of a liberty is based on the recognition that not all actions are the execution of a moral duty, and the claim that such "elbow room" is itself a moral good. The distinction between claim rights and liberties was embodied in the Roman distinction between *ius* and *dominum*.[4] Liberties will be very important to our discussion of the concept of appropriation, the process of creating ownable things from unowned things.

The existence of liberties leads Hohfeld to posit two additional forms of rights: **Powers** and **Immunities**. A power is the right to unilaterally act to change the rights of others. A good example is the assertion of copyright: by asserting my copyright, I automatically limit the liberties of other people to copy and sell my work. The obverse of a power is an immunity, which is a right *not* to have certain rights limited by the powers of another. I have the right to copy parts of a copyrighted work if my copying is "fair use" for personal or scholarly purposes. The doctrine of fair use thus creates an immunity from the power of a copyright holder.

Each of the four types of rights represents a paradigmatic kind of social relation, and carries a certain metaphorical baggage. The paradigm of a claim right is social obligation, with metaphors of responsibility, restraint, and charity. The paradigm of a liberty is personal freedom, with metaphors of creativity, autonomy, and privacy. The paradigm of a power is a law, with metaphors of control and organization. The paradigm of an immunity is citizenship, with metaphors of human dignity, privilege, and justice. All of these rights have both positive and negative connotations.

Property and Value

Before we discuss modes of ownership, we need to reflect briefly on the concept of value. There are two basic points to be made here.

First, we must make a distinction between private and social utility.

> Efficient organization entails the establishment of institutional arrangements and property rights that create an incentive to channel individual economic effort into activities that bring the private rate of return close to the social rate of return. The private rate of return is the sum of the net receipts which the economic unit receives from undertaking an activity. The social rate of return is the total net benefit (positive or negative) that society gains from the same activity. It is the private rate of return plus the net effect of the activity upon everyone else in the society.[5]

One time-honored method of getting people to create socially useful things is to invest them with property rights in their creations. Though we need to be realistic about incentives in what follows, we must not lose sight of the fact that economic incentives are neither necessary or sufficient conditions for the creation of social utility, especially when the work meets some private need of the author.

Private value can be divided into two categories: what we can call the *utility value* of a thing and what we can call its *value of tenure*. Utility value is the value of the thing in meeting some need. Value of tenure is value derived from the exclusive control of a thing. When something can't be shared (e.g., a coat), the existence of utility value depends on the value of tenure. Without exclusive control, the resource has no value. Such resources could be called "naturally private property." Resources where utility does not depend on value of tenure are potentially common. It may still be possible to impose value of tenure on it through exclusive control: this is equivalent to rent-seeking.[6]

Discussions of IP sometimes seem to assume that the only relevant value of a thing is its value of tenure. This assumption is seductive (since it makes economic analysis simple), but it is clearly false. As we shall see, authors have a variety of reasons for creating works other than exploiting their exclusive control of them. Value of tenure is important as a solution to market failure, but it still depends on utility value.

The concept of value is significant in what follows in two ways. First, it is the possibility that private value and social value can diverge that creates the public goods problem of IP. The possibility we need to consider here is whether or not a sufficiently large divergence (i.e., tremendous social value) could override private value. This kind of overriding is the basis of the principle of necessity in natural law theories: under some circumstances, the right to life can override property rights. [7]

A focus on social value suggests that we need to identify a non-personal source for the materials that constitute IP. This opens the possibility of introducing a new level of moral analysis. Classical natural law theories such as Locke's are driven by the need to show that appropriation from the commons is possible without infringing on the rights of the other commoners. However, there may be duties that take into account not only extant commoners, but future generations as well (similar to the concerns Rawls addresses with the principle of just saving).[8]

There is even the possibility that the commons itself might have intrinsic value that justifies restraints on the activities of commoners. Thus commoners might have duties to the commons that cannot be translated into rights held by the other extant commoners.

Ownership

We focus on *ownership* rather than *property* in our analysis because ownership is manifestly a *relation*, rather than a property of things. An attempt to construct a theory of property by starting with "properties" of things invites confusion. I concur with Tony Honoré when he writes "However it may be, it is clear that to stare at the meanings of the word 'thing' will not tell us what protected interests are conceived in terms of ownership."[9]

Honoré's analysis is an attempt to define universal features of ownership. "If ownership is provisionally defined as the *greatest possible interest in a thing that a mature legal system of law recognizes*, then it follows that, since all mature systems admit the existence of interests in things, all mature systems have a concept of ownership . . . In them certain legal incidents are common to different systems . . . Ownership and similar words stand not merely for the greatest interest in things in particular systems but for a type of interest with common features transcending particular systems."[10]

These features are best represented as cluster concepts: they are all involved in one or another form of ownership but do not individually constitute necessary or sufficient conditions for the applicability of the concept of ownership.

Honoré identifies eleven incidents[11] that characterize the concept of ownership, and uses the term "full liberal ownership" to refer to a system of ownership that contains all eleven simultaneously. Various subsets of the incidents can be used to define other forms of ownership, as we shall see below.

Control is at the heart of ownership. The incidents of ownership describe the rights associated with William Blackstone's classic definition of property as "that sole and despotic dominion which one man claims and exercises over the external things of the world, in total exclusion to of the right of any other individual in the universe."[12]

- The right to possess: "exclusive physical control of a thing, or to have such control as the nature of the thing admits."
- The right to the capital: "The right to the capital consists in the power to alienate the thing and the liberty to consume, waste, or destroy the whole or part of it."
- The right to use: "The right (liberty) to use the thing at one's discretion" where use is defined as "the owner's personal use and enjoyment of the thing owned."
- The right to manage: "The right to decide how and by whom the thing owned shall be used."
- The duty to prevent harm: "An owner's liberty to use and manage the thing owned as he chooses is subject to the condition that not only may he not use it to harm others, but he must prevent others using the thing to harm other members of society."
- The right to the income: income is "a surrogate of use, a benefit derived from forgoing the personal use of a thing and allowing others to use it for reward." The owner has a *prima facie* claim on economic value created by the non-consumptive use of the thing.
- Absence of term: "[the owner] should be able to look forward to remaining owner indefinitely if he so chooses and he remains solvent." The right to security presupposes that no agent in the legal system has the right to expropriate in a purely arbitrary manner. If such an agent (say, King Bob) existed, "ownership" would come to mean "ownership by King Bob."
- Transmissibility: "that the interest can be transmitted to the owner's successors, and so on ad infinitum."
- Residuary character: "the existence of B's lesser interest in a thing is clearly consistent with A owning it. To explain the usage in such cases

it is helpful to point out that it is a necessary but not sufficient condition of A's being owner that, either immediately or ultimately, the extinction of other interest would inure to his benefit."

- Liability to execution: "the liability of the owner's interest to be taken
  away from him for debt, either by the execution for a judgment debt or
  insolvency."[13]

Honoré's typology has a number of advantages. Different forms of ownership
can be characterized by different bundles of rights, without the need to specify
some property that is common to all forms of ownership. The typology also allows us to conceptually separate (a) control, (b) economic rights, and (c) the procedural aspect of property rights. This is extremely important, since the incidents
of control don't apply in a very natural way to IP. There is no meaningful way to
"fence in" a literary work in the way one can enclose a piece of real estate or a
pair of shoes. An author cannot "consume, waste, or destroy" an invention that
has been patented. It is also unclear how accountable authors should be held for
the uses others make of their ideas. However, the right to manage makes sense
for intellectual property, as does the right to the income and several of the incidents of duration.

In this work we will use the term "full ownership of intellectual property" to
mean the following set of fully realized incidents:

- The right to possess
- The right to use
- The right to manage
- The right to the income
- Residuary character
- Liability to execution

The following incidents apply to a limited extent:

- The right to the capital (there is no way to "consume or destroy" IP,
  though it can be alienated).
- Transmissibility (can be inherited, but not "ad infinitum").

The following incidents do not apply at all to IP:

- Absence of term (all forms of IP have limited term).
- Duty to prevent harm (the *modern* IP system does not function as a liability regime: authors are not generally held accountable for the uses
  that others might make of their works)

Notes

1. Charles Donahue, "Property Law," in *Encyclopedia Britannica*, 15th edition (Chicago: Encyclopedia Britannica, 1997), Volume 26, 180–205.

2. Wesley Newcomb Hohfeld, *Fundamental Legal Conceptions* (New Haven, CT: Yale University Press, 1919).

3. Becker, *Property Rights: Philosophical Foundations*, 8.

4. Tuck, *Natural Rights Theories*, 5–7.

5. North, *The Rise of the Western World*, 3.

6. See appendix A for a discussion of rent-seeking.

7. Tuck, *Natural Rights Theories*, 80.

8. Rawls, *A Theory of Justice*, 251–258.

9. Honoré, *Making Law Bind*, 181.

10. Honoré, *Making Law Bind*, 162.

11. In common–law terminology, an "incident" is a right that can be exercised at the discretion of its bearer (e.g., Hohfeld's powers and liberties).

12. William Blackstone, *Commentaries on the Laws of England*. Quoted in Mark Rose, *Authors and Owners*, (Cambridge, Mass: Harvard University Press, 1993), 7.

13. Honoré, *Making Law Bind*, 166–179.

Appendix C

The Statute of Anne

An act for the encouragement of learning, by vesting the copies of printed books in the authors or purchasers of such copies, during the times therein mentioned.

Whereas printers, booksellers, and other persons have of late frequently taken the liberty of printing, reprinting, and publishing, or causing to be printed, reprinted, and published, books and other writings, without the consent of the authors or proprietors of such books and writings, to their very great detriment, and too often to the ruin of them and their families: for preventing therefore such practices for the future, and for the encouragement of learned men to compose and write useful books; may it please your Majesty, that it may be enacted, and be it enacted by the Queen's most excellent majesty, by and with the advice and consent of the lords spiritual and temporal, and commons, in this present parliament assembled, and by the authority of the same;

That from and after the tenth day of April, one thousand seven hundred and ten, the author of any book or books already printed, who hath not transferred to any other the copy or copies of such book or books, share or shares thereof, or the bookseller or booksellers, printer or printers, or other person or persons, who hath or have purchased or acquired the copy or copies of any book or books, in order to print or reprint the same, shall have the sole right and liberty of printing such book and books for the term of one and twenty years, to commence from the said tenth day of April, and no longer; and

That the author of any book or books already composed, and not printed and published, or that shall hereafter be composed, and his assignee or assigns, shall have the sole liberty of printing and reprinting such book and books for the term of fourteen years, to commence from the day of the first publishing the same, and no longer; and

That if any other bookseller, printer or other person whatsoever, from and after the tenth day of April, one thousand seven hundred and ten, within the times

201

granted and limited by this act, as aforesaid, shall print, reprint, or import, or cause to be printed, reprinted, or imported, any such book or books, without the consent of the proprietor or proprietors thereof first had and obtained in writing, signed in the presence of two or more credible witnesses; or knowing the same to be so printed or reprinted, without the consent of the proprietors, shall sell, publish, or expose to sale, or cause to be sold, published, or exposed to sale, any such book or books, without such consent first had and obtained, as aforesaid: then such offender or offenders shall forfeit such book or books, and all and every sheet or sheets, being part of such book or books, to the proprietor or proprietors of the copy thereof, who shall forthwith damask, and make waste paper of them; and further,

That every such offender or offenders shall forfeit one penny for every sheet which shall be found in his, her, or their custody, either printed or printing, published, or exposed to sale, contrary to the true intent and meaning of this act; the one moiety thereof to the Queen's most excellent majesty, her heirs and successors, and the other moiety thereof to any person or persons that shall sue for the same, to be recovered in any of her Majesty's courts of record at Westminster, by action of debt, bill, plaint, or information, in which no wager of law, essoin, privilege, or protection, or more than one imparlance shall be allowed.

II. And whereas many persons may through ignorance offend against this act, unless some provision be made, whereby the property in every such book, as is intended by this act to be secured to the proprietor or proprietors thereof, may be ascertained, as likewise the consent of such proprietor or proprietors for the printing or reprinting of such book or books may from time to time be known;

Be it therefore further enacted by the authority aforesaid, that nothing in this act contained shall be construed to extend to subject any bookseller, printer, or other person whatsoever, to the forfeitures or penalties therein mentioned, for or by reason of the printing or reprinting of any book or books without such consent, as aforesaid, unless the title to the copy of such book or books hereafter published shall, before such publication, be entered in the register book of the company of Stationers, in such manner as hath been usual, which register book shall at all times be kept at the hall of the said company, and unless such consent of the proprietor or proprietors be in like manner entered as aforesaid, for every of which several entries, six pence shall be paid, and no more; which said register book may, at all seasonable and convenient time, be resorted to, and inspected by any bookseller, printer, or other person, for the purposes before-mentioned, without any fee or reward; and

The clerk of the said company of Stationers shall, when and as often as thereunto required, give a certificate under his hand of such entry or entries, and for every such certificate may take a fee not exceeding six pence.

III. Provided nevertheless, That if the clerk of the said company of Stationers for the time being, shall refuse or neglect to register, or make such entry or entries, or to give such certificate, being thereunto required by the author or proprietor of such copy or copies, in the presence of two or more credible witnesses, That then such person and persons so refusing, notice being first duly given of such refusal, by an advertisement in the Gazette, shall have the like benefit, as if such entry or entries, certificate or certificates had been duly made and given; and that the clerks so refusing, shall, for any such offense, forfeit to the proprietor of such copy or copies the sum of twenty pounds, to be recovered in any of her Majesty's courts of record at Westminster, by action of debt, bill, plaint, or information, in which no wager of law, essoin, privilege or protection, or more than one imparlance shall be allowed.

IV. Provided nevertheless, and it is hereby further enacted by the authority aforesaid, That if any bookseller or booksellers, printer or printers, shall, after the said five and twentieth day of March, one thousand seven hundred and ten, set a price upon, or sell, or expose to sale, any book or books at such a price or rate as shall be conceived by any person or persons to be too high and unreasonable;

It shall and may be lawful for any person or persons, to make complaint thereof to the lord archbishop of Canterbury for the time being, the lord chancellor, or lord keeper of the great seal of Great Britain for the time being, the lord bishop of London for the time being, the lord chief justice of the court of Queen's Bench, the lord chief justice of the court of Common Pleas, the lord chief baron of the court of Exchequer for the time being, the vice chancellors of the two universities for the time being, in that part of Great Britain called England; the lord president of the sessions for the time being, the lord chief justice general for the time being, the lord chief baron of the Exchequer for the time being, the rector of the college of Edinburgh for the time being, in that part of Great Britain called Scotland;

Who, or any one of them, shall and have hereby full power and authority, from time to time, to send for, summon, or call before him or them such bookseller or booksellers, printer or printers, and to examine and enquire of the reason of the dearness and inhauncement of the price or value of such book or books by him or them so sold or exposed to sale; and if upon such enquiry and examination it shall be found, that the price of such book or books is inhaunced, or any wise too high or unreasonable, then and in such case the said archbishop of Canterbury, lord chancellor or lord keeper, bishop of London, two chief justices, chief baron, vice chancellors of the universities, in that part of Great Britain called England, and the said lord president of the sessions, lord justice general, lord chief baron, and the rector of the college of Edinburgh, in that part of Great Britain called Scotland, or any one or more of them, so enquiring and examining, have hereby full power and authority to reform and redress the same, and to limit and settle the price of every such printed book and books, from time to time, according to the best of their judgments, and as to them shall seem just and reasonable; and

In case of alteration of the rate or price from what was set or demanded by such bookseller or booksellers, printer or printers, to award and order such bookseller and booksellers, printer and printers, to pay all the costs and charges that the person or persons so complaining shall be put unto, by reason of such complaint, and of the causing such rate or price to be so limited and settled; all which shall be done by the said archbishop of Canterbury, lord chancellor or lord keeper, bishop of London, two chief justices, chief baron, vice chancellors of the two universities, in that part of Great Britain called England, and the said lord president of the sessions, lord justice general, lord chief baron, and rector of the college of Edinburgh, in that part of Great Britain called Scotland, or any one of them, by writing under their hands and seals, and thereof publick notice shall be forthwith given by the said bookseller or booksellers, printer or printers, by an advertisement in the Gazette; and

If any bookseller or booksellers, printer or printers, shall, after such settlement made of the said rate and price, sell, or expose to sale, any book or books, at a higher or greater price, than what shall have been so limited and settled, as aforesaid, then, and in every such case such bookseller and booksellers, printer and printers, shall forfeit the sum of five pounds for every such book so by him, her, or them sold or exposed to sale; one moiety thereof to the Queen's most excellent majesty, her heirs and successors, and the other moiety to any person or persons that shall sue for the same, to be recovered, with costs of suit, in any of her Majesty's courts of record at Westminster, by action of debt, bill, plaint or information, in which no wager of law, essoin, privilege, or protection, or more than one imparlance shall be allowed.

V. Provided always, and it is hereby enacted, That nine copies of each book or books, upon the best paper, that from and after the said tenth day of April, one thousand seven hundred and ten, shall be printed and published, as aforesaid, or reprinted and published with additions, shall, by the printer and printers thereof, be delivered to the warehouse keeper of the said company of stationers for the time being, at the hall of the said company, before such publication made, for the use of the royal library, the libraries of the universities of Oxford and Cambridge, the libraries of the four universities in Scotland, the library of Sion College in London, and the library commonly called the library belonging to the faculty of advocates at Edinburgh respectively;

Which said warehouse keeper is hereby required within ten days after demand by the keepers of the respective libraries, or any person or persons by them or any of them authorized to demand the said copy, to deliver the same, for the use of the aforesaid libraries; and if any proprietor, bookseller, or printer, or the said warehouse keeper of the said company of stationers, shall not observe the direction of this act therein, that then he and they so making default in not delivering the said printed copies, as aforesaid, shall forfeit, besides the value of the said printed copies, the sum of five pounds for every copy not so delivered, as also the value of the said printed copy not so delivered, the same to be recovered by the Queen's majesty, her heirs and successors, and by the chancellor, masters, and scholars of any of the said universities, and by the president and fellows of

Sion College, and the said faculty of advocates at Edinburgh, with their full costs respectively.

VI. Provided always, and be it further enacted, That if any person or persons incur the penalties contained in this act, in that part of Great Britain called Scotland, they shall be recoverable by any action before the court of session there.

VII. Provided, That nothing in this act contained, do extend, or shall be construed to extend to prohibit the importation, vending, or selling of any books in Greek, Latin, or any other foreign language printed beyond the seas; any thing in this act contained to the contrary notwithstanding.

VIII. And be it further enacted by the authority aforesaid, That if any action or suit shall be commenced or brought against any person or persons whatsoever, for doing or causing to be done any thing in pursuance of this act, the defendants in such action may plead the general issue, and give the special matter in evidence; and if upon such action a verdict be given for the defendant, or the plaintiff become nonsuited, or discontinue his action, then the defendant shall have and recover his full costs, for which he shall have the same remedy as a defendant in any case by law hath.

IX. Provided, That nothing in this act contained shall extend, or be construed to extend, either to prejudice or confirm any right that the said universities, or any of them, or any person or persons have, or claim to have, to the printing or reprinting any book or copy already printed, or hereafter to be printed.

X. Provided nevertheless, That all actions, suits, bills, indictments or informations for any offence that shall be committed against this act, shall be brought, sued, and commenced within three months next after such offence committed, or else the same shall be void and of none effect.

XI. Provided always, That after the expiration of the said term of fourteen years, the sole right of printing or disposing of copies shall return to the authors thereof, if they are then living, for another term of fourteen years.

Bibliography

Alfino, Mark. "Intellectual Property and Copyright Ethics." *Business and Professional Ethics Journal* 10, no. 2 (1991): 85–109.

Andrews, Lori, and Dorothy Nelkin. *Body Bazaar: The Market for Human Tissue in the Biotechnology Age.* Westminister, MD: Crown Publishing, 2001.

Arendt, Hannah. *The Human Condition.* Chicago: University of Chicago Press, 1958.

Ashcraft, Richard. *Revolutionary Politics and Locke's Two Treatises of Government.* Princeton, NJ: Princeton University Press, 1986.

Barlow, John Perry. "Coming Into the Country." *Communications of the ACM* 34 no. 3, (1991): 19–21.

————. "Selling Wine Without Bottles: The Economy of Mind on the Global Net." In Ludlow, *High Noon*, 9–34.

Baslar, Kemal. *The Concept of the Common Heritage of Mankind in International Law.* Norwell, MA: Martinus Nijhoff, 1998.

Bayles, Michael D. "Brand Name Extortionists, Intellectual Prostitutes, and Generic Free Riders." *International Journal of Applied Philosophy* 2, no. 3 (1984): 13–26.

Becker, Lawrence C. *Property Rights: Philosophical Foundations.* New York: Routledge Kegan Paul, 1977.

Benkler, Yochai. "Free as the Air to Common Use: First Amendment Constraints on Enclosure of the Public Domain." *New York University Law Review* 74 (May 1999): 354.

Bontchev, Vessilin. "The Bulgarian and Soviet Virus Factories." 1991 paper. www.texfiles.com/virus/bulgfact.txt (accessed April 12, 2004).

Boyle, James. "Intellectual Property Online: A Young Person's Guide." *Journal of Law and Technology*, 1995.

————. *Shamans, Software, and Spleens: Law and the Construction of the Information Society.* Cambridge, MA: Harvard University Press, 1996.

————. "Foucault In Cyberspace: Survelliance, Sovereignty and Hardwired Censors." www.wcl. american.edu/pub/faculty/boyle/papers/foucault.htm, 1997.

Boyle, James, and Bruce A. Lehman. 1996. "The Debate on the NII White Paper." Correspondence between Bruce Lehman and James Boyle. Posted on the WWW by Boyle.

Braudel, Fernand. *Civilization and Capitalism 1500–1800.* New York: Harper and Row, 1981.

Brennan, Geoffrey. "Economics." In Goodin and Pettit, *Contemporary Politicial Philosophy*, Oxford: Basil Blackwell, 1997, 123–156.

Breyer, Stephen. "The Uneasy Case for Copyright: A Study of Copyright in Books, Photocopies, and Computer Programs." *Harvard Law Review* 84 (1970): 281.

Bringsjord, Selmer. "In Defense of Copying." *Public Affairs Quarterly* 3, no. 1 (1989): 1–9.

Buckle, Stephen. *Natural Law and the Theory of Property: Grotius to Hume.* New York: Cambridge University Press, 1991.

Bynum, Terrel Ward. "Introduction and Overview: Global Information Ethics." *Science and Engineering Ethics* 2 (1996): 131–136.

Carey, David. "Should Computer Programs be Ownable?" *Metaphilosophy* 24 (1993): 76–84.

———. "A Reply to Johnson's Reply to 'Should Computer Programs Be Ownable?'" *Metaphilosophy* 24 (1993): 91–96.

Carpenter, Stanley R. "Sustainability and Common Pool Resources: Alternatives to Tragedy." *Philosophy and Technology* 3, no. 4 (1998): 36–57.

Castells, Manuel. *The Rise of the Network Society.* London: Blackwell, 1996.

Caulfield, Timothy, and Bryn Williams-Jones, eds. *The Commercialization of Genetic Research: Ethical, Legal, and Policy Issues.* Dodrecht: Kluwer, 1999.

CEC. "Proposal for a European Parliment and Council Directive on the Harmonization of Certain Aspects of Copyright and Related Rights in the Information Society." Technical Report, European Union, 1997.

Cohen, Julie E. "A Right to Read Anonymously: A Closer Look at Copyright Management in Cyberspace." *Connecticut Law Review* 28 (1996): 981.

Crespi, R. Stephen. "The Case for Patenting Biotechnological Inventions." In Sigrid Sterckx ed., *Biotechnology, Patents, and Morality.* Brookfield, VT: Ashgate, 2000. Chapter 26, 277–296.

Davies, Kevin. *Cracking the Genome: Inside the Race to Unlock Human DNA.* New York: Free Press, 2001.

Davis, Randall, Pamela Samuelson, Mitchell Kapor, and J. H. Reichman. "A New View of Intellectual Property and Software." *Communications of the ACM* 39, no. 3 (1996): 21–30.

DePalma, Anthony. "The Slippery Slope of Patenting Farmer's Crops." *New York Times*, May 24, 2000.

Desowitz, Robert. *The Malaria Capers.* New York: Norton, 1991.

Dibona, Chris, ed. *Open Sources: Voices from the Open Source Revolution.* San Francisco: O'Reilly Associates, 1999.

Donahue, Charles. "Property Law." In *Encyclopedia Britannica*, Volume 26, 15th, 180–205. Chicago: Encyclopedia Britannica, 1997.

Drahos, Peter, ed. *Intellectual Property.* International Library of Essays in Law and Legal Theory. Second Series. Brookfield, VT: Ashgate Dartmouth, 1999.

———. "Information Feudalism in the Information Society." In Peter Drahos, ed., *Intellectual Property.* Brookfield, VT: Ashgate Dartmouth, 1999, 471–484.

———. "Global Property Rights in Information: The Story of TRIPS at GATT." In Peter Drahos, ed., *Intellectual Property.* Brookfield, VT: Ashgate Dartmouth, 1999, 419–432.

Dreger, Alice Domurat. "Metaphors of Morality in the Human Genome Project." In Phillip Sloan, ed., *Controlling Our Destinies: Historical, Philosophical, Ethical, and Theological Perspectives on the Human Genome Project.* (Notre Dame IN: Notre Dame Press, 2001), 155–184.

DSF. "What is the Campaign for Access to Essential Medicines?" www.doctorswithoutborders.org/ advocacy/access/crisis.html.

Eisenstein, Elizabeth L. *The Printing Press as an Agent of Change.* Cambridge University Press, 1979.

Elias, Stephen. *Patent, Copyright & Trademark.* Berkeley, CA: Nolo Press, 1999.

Elton, Charles I. "Early Forms of Landholding," 1886. www.socsci. mcmaster.ca~/econ/ugcm/3113/~misc/elton.html.

Fisher, William. "The Growth of Intellectual Property: A History of the Ownership of Ideas in the United States." www.law.harvard.edu/Academic_Affairs/coursepages/tfisher/iphistory.html.

Foucault, Michel. "What is an Author?" In *The Foucault Reader*, edited by Paul Rabinow. New York: Pantheon, 1969.

——. *The Order of Things.* New York: Random House, 1970.

Frank, Robert, and Philip Cook. *The Winner Take All Society: Why the Few at the Top Get so Much More than the Rest of Us.* New York: Penguin, 1995.

Geller, Paul Edward. "Must Copyright be For Ever Caught Between Marketplace and Authorship Norms?" In Sherman and Strowel, eds., *Of Authors and Origins: Essays on Copyright Law*, (New York: Oxford University Press, 1994), 159–201.

——. "The Universal Electronic Archive: Issues in International Copyright." *International Review of Industrial Property and Copyright Law* 25 (1994): 54–69.

Genewatch. "Privatising Knowledge, Patenting Genes." Briefing Number 11 - PDF file, http://www.genewatch.org.

Gibson, William. *Mona Lisa Overdrive.* New York: Bantam, 1988.

——. *Idoru.* New York: G.P. Putnam's Sons, 1996.

Ginsburg, Jane C. "Copyright Without Walls?: Speculations on Literary Property in the Library of the Future." *Representations* 42 (1993): 53.

——. "A Tale of Two Copyrights: Literary Property in Revolutionary France and America." In Sherman and Strowel, eds., *Of Authors and Origins: Essays on Copyright Law* (New York: Oxford University Press, 1994) 131–158.

——. "Putting Cars on the 'Information Superhighway': Authors, Exploiters and Copyright in Cyberspace." *Columbia Law Review* 95 (1995): 1466.

Glass, Richard S, and Wallace A Wood. "Situational Determinants of Software Piracy: An Equity Theory Perspective." *Journal of Business Ethics* 15 (1996): 1189–1198.

Goldstein, Paul. "Copyright." *Journal of the Copyright Society of the U.S.A.* (1991), 109–110.

——. *Copyright's Highway: From Gutenberg to the Celestial Jukebox.* New York: Hill and Wang, 1996.

Goodin, Robert E., and Phillip Pettit, eds. *A Companion to Contemporary Political Philosophy.* Blackwell Companions to Philosophy. Oxford: Basil Blackwell, 1995.

Gordon, Wendy. "A Property Right in Self-expression: Equality and Individualism in the Natural Law of Intellectual Property." *Yale Law Journal* 102 (1993): 1540–1578.

——. "Touring the Certainties of Property and Restitution: A Journey to Copyright and Parody." in E. Mackaay, ed., *Certitudes Du Droit/Certainty and the Law* (Montreal: Themis Publishing, 2000).

——. "Authors, Publishers, and Public Goods: Trading Gold for Dross," *Loyola of Los Angeles Law Review* 36 (2002): 159.

Grahame, Kenneth. *The Wind in the Willows.* Farmington Hills, MI: Charles Scribner, 1960.

Greaves, Tom, ed. *Intellectual Property Rights for Indigenous Peoples: A Sourcebook.* Oklahoma City: The Society for Applied Anthropology, 1994.

Grotius, Hugo. *On the Freedom of the Seas.* New York: Oxford University Press, 1917.

———. *On the Laws of War and Peace.* Oxford University Press, 1923.

Hardin, Garrett. "The Tragedy of the Commons." *Science* 162 (1968):1243–1248.

Heckel, Paul. "Debunking the Software Patent Myths." In Peter Ludlow ed., *High Noon on the Electronic Frontier* (Cambridge, MA: MIT Press, 1997), 63-108.

Heller, Michael, and Rebecca Eisenberg. "Can Patents Deter Innovation?: The Anticommons in Biomedical Research." *Science* 280 (1998): 698–701.

Hesse, Carla. *Publishing and Cultural Politics in Revolutionary Paris, 1789–1810.* Volume 12 of Studies on the History of Society and Culture. Los Angeles, CA: University of California Press, 1991.

Hettinger, Edwin C. "Justifying Intellectual Property." *Philosophy and Public Affairs* 18, no. 1 (1989): 31–52.

Hill, Christopher. *The World Turned Upside Down: Radical Ideas during the English Revolution.* New York: Penguin Books, 1972.

———. *Winstanley: The Law of Freedom and Other Writings.* New York: Cambridge University Press, 1973.

———. *The Century of Revolution: 1603-1714.* Norton Library History of England. New York: W.W. Norton, 1980.

Hobbes, Thomas. *On the Citizen.* Cambridge Texts in the History of Political Thought. New York: Cambridge University Press, 2001.

———. *Leviathan.* Volume 21 of *Great Books of the Western World.* Chicago: Encyclopedia Brittanica, Inc., 1991.

Hobsbawn, E. J. *Industry and Empire.* Volume 3 of *The Penguin Economic History of Britain.* New York: Penguin, 1968.

Hohfeld, Wesley Newcomb. *Fundamental Legal Conceptions.* New Haven, CT: Yale University Press, 1919.

Holderness, B. A. *Pre-industrial England: Economy and Society from 1500 to 1750.* Lanham, MD: Rowman and Littlefield, 1976.

Honoré, Tony. "Ownership." In *Making Law Bind.* Oxford: Oxford University Press, 1987.

Hughes, Justin. "The Philosophy of Intellectual Property." *Georgetown Law Journal* 77 (1988): 287.

Johns, Adrian. *The Nature of the Book.* Chicago: University of Chicago Press, 1998.

Johnson, Deborah G. "Should Computer Programs be Owned?" *Metaphilosophy* 16 (1985): 276–288.

———. "Equal Access to Computing, Computing Expertise and Decision Making about Computers." *Business and Professional Ethics Journal* 4 (1985): 95–104.

———. "A Reply to 'Should Computer Programs be Ownable?'" *Metaphilosophy* 24 (1993): 85–90.

Johnson, Deborah, and Helen Nissenbaum, eds. *Computers, Ethics and Social Values.* Paramus, NJ: Prentice Hall, 1995.

Kaplan, Benjamin. *An Unhurried View of Copyright.* New York: Columbia University Press, 1967.

Karjala, Dennis S. "Judicial Review of Copyright Term Extension Legislation." *Loyola of Los Angeles Law Review* 36 (2002): 199–251.

Kilcullen, John. 1995. "The Origin of Property: Ockham, Grotius, Pufendorf, and Some Others." www.humanities.mq.edu.au/Ockham/wpr.html.

Knoppers, Bartha Maria. 1999. "Sovereignty and Sharing." In Caulfield and Williams-Jones, eds. *The Commercialization of Genetic Research: Ethical, Legal, and Policy Issues.* Dodrecht: Kluwer,1999. Chapter 1, 1–12.

Kuflik, Arthur. "Moral Foundations of Intellectual Property Rights." In Johnson and Nissenbaum , eds. *Computers, Ethics and Social Values* (Paramus, NJ: Prentice Hall, 1995). 169–180.

Lange, David. "Recognizing The Public Domain." *Law and Comtemporary Problems* 44, no. 4 (1981): 147–178.

Lehman, Bruce. Software Patents Town Meeting. Acquired from Electronic Frontiers Foundation Online Archive. Verbatim transcript of testimony given to Commissioner of Patents Bruce Lehman in San Jose, CA 1/26/94–1/27/94.

Lehman, Bruce A. *Final Report of the Working Group on Intellectual Property Rights* (known as the "NII White Paper.") Technical Report, National Information Infrastructure Task Force. Published by the U.S. Patent and Trademark Office, 1994.

Lessig, Lawrence. "The Path of Cyberlaw." *Yale Law Journal* 104 (May 1995): 1403.

Levy, Steven. *Crypto.* New York: Random House, 2000.

Lewontin, R. C. "People Are Not Commodities." *New York Times*, January 23, 1999.

Litman, Jessica. "A Rose By Any Other Name: Computer Programs and the Idea/Expression Boundary." *Emory Law Journal* 34 (1985): 741.

————. "The Public Domain." *Emory Law Journal* 39 (1990): 965–1023.

————. "The Exclusive Right to Read." *Cardozo Arts and Entertainment Law Journal* 13 (1994): 29.

Library of Congress. "The Copyright Law of the United States." Circular 92, Library of Congress Copyright Office, 1998.

————. "Duration of Copyright." Circular 15a, Library of Congress Copyright Office, April 1999.

Locke, John. *Two Treatises of Government.* Cambridge Texts in the History of Political Thought. New York: Cambridge University Press, 1988.

Ludlow, Peter, ed. *High Noon at the Electronic Frontier.* Cambridge, MA: MIT Press, 1997.

MacLeod, Christine. *Inventing the Industrial Revolution: the English Patent System, 1660-1800.* New York: Cambridge University Press, 1988.

MacPherson, C. B. *The Political Theory of Possessive Individualism.* Oxford: Oxford University Press, 1962.

Martin, Brian. "Against Intellectual Property." *Philosophy and Social Action* 21 no. 3 (1995): 7–22.

Marx, Karl. *Capital.* Volume 50 of *Great Books of the Western World.* Edited by Friedrich Engels. Chicago: Encyclopedia Brittanica, Inc. Translated from the third German edition by Samuel Moore and Edward Aveling.

McNeil, Donald G. "Profits on Cosmetic Save a Cure for Sleeping Sickness." *New York Times*, February 9, 2001.

Mell, Patricia. "Seeking Shade in a Land of Perpetual Sunlight: Privacy as Property in the Electronic Wilderness." *Berkeley Technology Law Journal* 11 (1996): no. 1.

Merges, Robert, and Richard Nelson. "On the Complex Economics of Patent Scope." *Columbia Law Review* 90 (1990): no. 4.

Merges, Robert P. "Property Rights Theories and the Commons: The Case of Scientific Research." In *Scientific Innovation, Philosophy, and Public Policy*, edited by Ellen

Paul, Fred Miller, and Jeffery Paul, 145–167. New York: Cambridge University Press, 1996.

Merges, Robert. "Institutions for Intellectual Property Transactions: The Case of Patent Pools." www. law.berkeley.edu/institutes/bclt/pubs/merges/pools.pdf

Milton, John. *Areopagitica: A Speech for the Liberty of Unlicensed Printing* Volume 29 of *Great Books of the Western World*. Chicago: Encyclopedia Britannica, 1991.

Moglen, Eben. "Anarchism Triumphant: Free Software and the Death of Copyright." *First Monday* 4 (1999): no. 8.

More, Thomas. *Utopia*. Norton Critical Editions. New York: Norton, 1975. Translated and edited by Robert Merrihew Adams.

Moykr, Joel. *The Lever of Riches: Technological Creativity and Economic Progress*. New York: Oxford University Press, 1990.

Munzer, Stephen R. *A Theory of Property*. Cambridge Studies in Philosophy and Law. New York: Cambridge University Press, 1990.

Netanal, Neil W. "Copyright and a Democratic Civil Society." *Yale Law Review* 106 (1996): 283.

Nissenbaum, Helen. "Should I Copy My Neighbor's Software?" In Johnson and Nissenbaum, *Computers, Ethics, and Social Values* (Paramus, NJ: Prentice Hall, 1995), 201–212.

———. "Accountability in a Computerized Society." *Science and Engineering Ethics* 2 (1996): 25–42.

North, Douglass C., and Robert P. Thomas. *The Rise of the Western World: a New Economic History*. New York: Cambridge University Press, 1973.

Nozick, Robert. *Anarchy, State and Utopia*. New York: Basic Books, 1974.

William of Ockham. *A Short Discourse on the Tyrannical Government Over Things Divine and Human*. Cambridge Texts in the History of Political Thought. Cambridge: Cambridge University Press, 1992.

Olivecrona, Karl. "Locke's theory of Appropriation." *Philosophical Quarterly* 24 (1974): 220–234.

Ostrom, Elinor. *Governing the Commons: The Evolution of Institutions for Collective Action*. The Political Economy of Institutions and Decisions. New York: Cambridge University Press, 1990.

Paine, Lynn Sharp. "Trade Secrets and the Justification of Intellectual Property." *Philosophy and Public Policy* 20 (1991): 247.

Parisi, Francesco, Norbert Schulz, and Ben Depoorter. "Duality in Property: Commons and AntiCommons." Working paper 00-16, University of Virginia School of Law, 2000.

Patterson, L. Ray. *Copyright in Historical Perspective*. Nashville, TN: Vanderbilt University Press, 1968.

Patterson, L. Ray, and Stanley Lindberg. *The Nature of Copyright: A Law of User Rights*. Athens, GA: University of Georgia Press, 1991.

Paul, Jeffrey, ed. *Reading Nozick: Essays on Anarchy, State and Utopia*. Lanham, MD: Rowman and Littlefield, 1981.

Pettit, Phillip. "Analytical Philosophy." In Goodin and Pettit, eds. *Contemporary Political Philosophy* (Oxford: Basil Blackwell, 1995), 7–38.

Posner, Richard. "The Economic Approach to Law." In David Adams, ed. *Philosophical Problems in the Law* (Belmont, CA: Wadsworth, 2001), 116–126.

Pressman, David. *Patent It Yourself*. Berkeley, CA: Nolo Press, 1999.

Price, William Hyde. *The English Patents of Monopoly*. Volume 1 of *Harvard Economic Studies*. Cambridge, MA: Harvard University Press, 1913.

Pufendorf, Samuel. *The Political Writings of Samuel Pufendorf.* Oxford: Oxford University Press, 1995.

Radin, Margeret Jane. "Property and Personhood" *Stanford Law Review* 34 (1982): 957. cyber.law.harvard.edu/ipcoop/82radi.html.

Rand, Ayn. "Patents and Copyrights." *Objectivist Newsletter* 3 (1964), no. 5.

Rawls, John. *A Theory of Justice.* Cambridge, MA: Harvard University Press, 1971.

Reidenberg, Joel. "Lex Informatica: The Formulation of Information Policy Rules through Technology." Forthcoming in *Texas Law Review.*

Rivette, Kevin G., and David Kline. "Discovering New Value in Intellectual Property." *Harvard Business Review* 78 (2000), no. 1: 54–66.

Rose, Carol A. *Property and Persuasion: Essays on the History, Theory and Rhetoric of Ownership.* New Perspectives on Law, Culture and Society. Boulder, CO: Westview Press, 1994.

———. "Expanding the Choices for the Global Commons: Comparing Newfangled Tradable Allowance Schemes to Old-fashioned Common Property Regimes." *Duke Environmental Law and Policy Forum* 10 (2000): 45–72.

Rose, Mark. *Authors and Owners.* Cambridge, MA: Harvard University Press, 1993.

———. "The Author as Proprietor: Donaldson v. Becket and the Geneology of Modern Authorship." In Sherman and Strowel, eds., *Of Authors and Origins: Essays on Copyright Law* (New York: Oxford University Press, 1994), 23–55.

Rousseau, Jean-Jacques. "A Discourse on the Origin of Inequality." Volume 35 of *Great Books of the Western World.* Chicago: Encyclopedia Brittanica Inc., 1991.

———. *The Social Contract.* Volume 35 of *Great Books of the Western World.* Chicago: Encyclopedia Brittanica Inc., 1991.

Ryan, Alan. *Property and Political Theory.* Oxford: Basil Blackwell, 1984.

Sabine, George H., ed. *The Collected Works of Gerrard Winstanley.* Ithaca, NY: Cornell University Press, 1941.

Samuels, Edward. *The Illustrated History of Copyright.* New York: Thomas Dunne Books: St. Martins Press, 2000.

———. "No. 01-618, Eldred v. Ashcroft: Brief of Amicus Curiae Edward Samuels in support of Respondent." www.edwardsamuels.com/copyright/beyond/cases/eldredamicus.htm, July 2002.

Samuelson, Pamela. "Is Information Property?" *Communications of the ACM* 34 no. 3 (March 1991): 15.

———. "The Copyright Grab." *Wired* 4, no. 1 (Jan 1996).

Sandburg, Brenda. "Battling the Patent Trolls." *The Recorder*, 2001.

Schatz, Ulrich. "Patents and Morality." In Sigrid Sterckx, ed., *Biotechnology, Patents, and Morality.* Brookfield, VT: Ashgate, 2000, Chapter 14.

Schlachter, Eric. "The Intellectual Property Renaissance in Cyberspace: Why Copyright Law Could be Unimportant on the Internet." *Berkeley Technology Law Journal* 12 (1997), no. 1.

Schneewind, J. B. *The Invention of Autonomy: A History of Modern Moral Philosophy.* New York: Cambridge University Press, 1998.

Sen, Amartya. "Rational Fools." *Philosophy and Public Affairs,* vol. 6 (1977): 314–344.

———. *On Ethics and Economics.* Oxford: Basil Blackwell, 1987. 1986 Royer lectures, University of California at Berkeley.

Sherman, Brad, and Alain Strowel, eds. *Of Authors and Origins: Essays on Copyright Law.* New York: Oxford University Press, 1994.

Shattuck, Roger. *Forbidden Knowledge: From Prometheus to Pornography.* New York: St. Martin's Press, 1996.

Shiva, Vandana. *Biopiracy: The Plunder of Nature and Knowledge*. New York: East End Press, 1995.

———. *Protect or Plunder: Understanding Intellectual Property Rights*. Global Issues in a Changing World. London: Zed Books, 2001.

Shulman, Seth. "Cashing In on Medical Patents." MIT Technology Review 101, no. 2 (February 1998): 38–45.

———. *Owning The Future*. Boston: Houghton Mifflin, 1999.

Sloan, Phillip, ed. *Controlling Our Destinies: Historical, Philosophical, Ethical, and Theological Perspectives on the Human Genome Project*. Notre Dame, IN: Notre Dame Press, 2001.

Snyder, Solomon. *Drugs and the Mind*. New York: Scientific American Library, 1986.

Spector, Horacio. "IP Skepticism." *International Journal of Applied Philosophy* 6 no. 2 (1991): 65–67.

Sreenivasan, Gopal. *The Limits of Lockean Rights in Property*. Oxford: Oxford University Press, 1995.

Stallman, Richard. "The GNU Manifesto." www.gnu.org, 1985.

———. "Against User Interface Copyright." February 1991. Acquired from Electronic Frontiers Foundation Online Archive.

———. "Why Software Should be Free." In Johnson and Nissenbaum, eds., *Computers, Ethics, and Social Values* (Paramus, NJ: Prentice Hall, 1995), 190–200.

Stefik, Mark, ed. *Internet Dreams: Archetypes, Myths and Metaphors*. Cambridge, MA: MIT Press, 1996.

———. "Shifting the Possible: How Trusted Systems and Digital Property Rights Challenge Us to Rethink Digital Publishing." *Berkeley High Technology Law Journal* 12 (1997), no. 1.

Steidlmeier, Paul. "The Moral Legitimacy of Intellectual Property Claims: American Business and Developing Country Perspectives." *Journal of Business Ethics* 12 (1993): 157–164.

Stein, Lincoln D. "Napster: Asking for Trouble. Getting It." *Web Techniques* 5, no. 5 (May 2000): 12–15.

Sterckx, Sigrid, ed. *Biotechnology, Patents, and Morality*. Brookfield, VT: Ashgate, 2000.

Stevenson, Glenn G. *Common Property Economics: A General Theory and Land Use Applications*. New York: Cambridge University Press, 1991.

Stix, Gary. 2001. "A License for Copycats?" *Scientific American* 284, no. 7 (July 2001): 36.

Suber, Peter. "What is Software?" *Journal of Speculative Philosophy* 2 (1988): 89–119.

Swinyard, W. R., H. Rinne, and A. Keng Kau. "The Morality of Software Piracy: A Cross-cultural Analysis." *Journal of Business Ethics* 9, no. 8 (1990): 655–664.

Tallmo, Karl-Erik, *The History of Copyright: A Critical Overview with Source Texts in Five Languages*. (forthcoming from nisus press) www.copyrighthistory.com

Thirsk, Joan, and J. P. Cooper, eds. *Seventeenth-century Economic Documents*. Oxford: Oxford University Press, 1972.

Torvalds, Linus, and David Diamond. *Just for Fun: The Story of an Accidental Revolutionary*. New York: Harper Business, 2001.

Tuck, Richard. *Natural Rights Theories: Their Origin and Development*. New York: Cambridge University Press, 1979.

———. *Philosophy and Government, 1572–1651*. New York: Cambridge University Press, 1993.

Tully, James. *A Discourse on Property: John Locke and His Adversaries.* New York: Cambridge University Press, 1980.

———. "Rediscovering America: The *Two Treatises* and Aboriginal Rights." Chapter 5 of *An Approach to Political Philosophy: Locke in Contexts.* New York: Cambridge University Press, 1993, 137–176.

U.S. Patent and Trademark Office. "Basic Facts About Registering a Trademark." Technical Report, 1992.

Waldron, Jeremy. "Enough and as Good Left For Others." *Philosophical Quarterly* 29 (1979): 319–328.

———. *The Right to Private Property.* New York: Oxford University Press, 1989.

Warren, Samuel, and Louis D. Brandeis. "The Right to Privacy." *Harvard Law Review* 4 (1890): 193.

Warshofsky, Fred. *Patent Wars: The Battle to Own the World's Technology.* Hoboken, NJ: John Wiley and Sons, 1994.

Wayner, Peter. *Free for All: How Linux and the Free Software Movement Undercut the High-Tech Titans.* New York: HarperCollins, 2000.

Yen, Alfred C. "Restoring the Natural Law: Copyright as Labor and Possession." *Ohio State Law Journal* 51 (1990): 517–559.

Index

active media, 52, 53, 56
Addison, Joseph, 36, 181
anticommons, 19, 54, 77, 98, 99,
    107, 155, 162, 165, 168, 189-
    191
appropriation:
    definition, 72
Aquinas, Thomas, 67
Arendt, Hannah, 89, 90, 96
ASCAP, 55
author's rights:
    argument from analogy, 35,
    36
authorship:
    as appropriation from a com-
    mons, 2; moral rights, 16
    romantic theory of, 14-16,
    52, 96, 147, 158

Bacon, Francis, 47
Baslar, Kemal, 160-162
Bayh-Dole Act, 168
Becker, Wesley, 2, 194
Bentham, Jeremy, 126
Berne Convention, 40
Blackstone, William, 127, 197
Bleistein v. Donaldson
    Lithographing Company, 51,
    132
BMI, 55
Bontchev, Vessilin, 120
Boyle, James, 2, 109, 158, 159,
    174, 175, 179
Bringsjord, Selmer, 54

Chakrabarty, Ananda, 155
Common Heritage of Mankind,
    160-163

commons:
    as solution for distributive
    justice, 67; as third form of
    property, 68; can be artificial,
    88; conquistador problem,
    70, 78, 80-82, 84, 101, 153,
    155, 164; consumptive use,
    73; definition, 71; egalitarian
    nature of, 67; external prob-
    lems, 67, 69; forms of use,
    71; frontier, 72; internal
    problems, 67, 70; problem of
    anticommons, 19; problem of
    artificiality, 70; problem of
    consent, 70; problem of late-
    comers, 70; problem of
    scarcity, 70; requirement for
    persistence, 75; strict, 72
    theoretical advantages, 69;
    third form of property, 19,
    69; unconfinable entities, 75
compulsory licensing, 52, 54, 55,
    167, 185
computer software, 56-58
    copy protection, 58, 59, 121
conquistador problem, 70, 78, 80,
    82, 84, 101, 153, 155, 164
copyright:
    1976 Copyright Act, 40, 56,
    57, 133; broadcasting, 2, 52,
    54-56, 61, 91; development
    of printing, 48; expanded
    scope, 26; fair use, 27, 28,
    61, 62; fixation replaces writ-
    ing, 56; formalities, 26, 90,
    100; French, 55; idea vs. ex-
    pression, 26-28, 54, 87, 89,
    91, 183, 185-187, 198; im-
    possible to prove orginality,

217

149; performance rights, 2,
26; performances, 55, 125;
perpetual common-law, 36;
photographs, 26, 40, 51-54;
recording technology, 26, 40,
51-56, 125; role of physical
constraints, 62; sheet music,
26, 53, 136
Copyright Term Extension Act,
40, 41, 43; increase in copyright term, 4; *quid pro quo*
argument, 41-43; retroactive
extension of term, 41, 42
cybersquatting, 148

debts between generations, 93
deCODE Genetics, 99-101
Defoe, Daniel, 2, 36, 142
Diamond v. Chakrabarty, 155
Diamond v. Diehr, 57
Digital Millennium Copyright Act,
41, 58, 63
Disney Corporation, 40
Donahue, Charles, 193
Donaldson v. Becket, 33, 37, 42
analogy with patents, 38; ar-
gument against common-law
right, 37, 38; other unprotect-
ed creativity, 39
Dworkin, Ronald, 127

econodwarves, 130, 184
Eldred v. Ashcroft, 33, 39, 41-44
enclosure:
definition, 72
encryption software, 59
English Patents of Monopoly, 4,
47, 48, 136, 188; negative effects, 47; reforms of 1624, 48

fair use, 63, 106, 125, 133, 190,
195
Federal Technology Transfer Act,
168
Filmer, Robert, 74
Foucault, Michel, 2, 90, 139-141
Francis of Assisi, 110
Franciscan poverty, 110, 112
free riders, 6, 15
Free Software Movement, 20, 79,

110, 119, 120
full liberal ownership, 10, 21, 25,
34, 69, 71, 84, 141, 147, 176,
197

genes:
as novel, 157; discovered vs.
invented, 156; patents, 156
geographic metaphor, 95
Ginsberg, Ruth Bader, 41
Goldstein, Robert, 14, 15, 20
Gordon,Wendy, 17
Grotius, Hugo, 3, 80-82, 94, 109,
110, 112-114, 121, 166, 177

Hardin, Garrett, 77, 78
Harvard Oncomouse, 164
Harvard University, 164
Hayflick, Leonard, 99
Heidegger, Martin, 89
Hobbes, Thomas, 79, 80, 127, 128
Hohfeld, Wesley Newcomb, 194
Holmes, Oliver Wendell Jr., 51,
132
Honoré, Tony, 141, 196-198
Hughes, Justin, 88, 92, 143-145

Iceland, 99-101
ideas:
as independent of authors,
90; unownable character, 91
intellectual commons:
barriers to ownership, 87; dualities with physical commons, 98; intuitive rationales,
19; need for replenishment,
22; zones of, 87
intellectual property (IP):
commodification of, 52, 159,
160; concept vs conception, 3
definition, 1; differences
from physical property, 5;
ethical constraints, 3, 4; incentive theory, 42; instrumentalist objection, 8, 9;
limitation of term, 1, 7, 8, 89,
93, 173, 174; scarcity and,
17; technological change and,
5, 6

intellectual property rights:
author-centered theory, 2, 5,
10, 14, 137, 139, 141, 145-
149, 173, 177; deep ecology,
20; deep ecology justification, 182; monolithic, 5;
publisher-centered, 5; role of
incentives, 9, 10; user-centered theory, 5; utilitarian
theory, 5

Jefferson, Thomas, 16, 17
Johns, Adrian, 35

Kaplan, Bernard, 2
Knoppers, Bartha Maria, 163

letters patent, 47
Lewontin, R. C., 101
libertarianism, 19, 75, 126
Linux, 20
Litman, Jessica, 2, 9, 14, 92
Locke, John, 82, 83, 93, 109, 116,
117, 143, 177, 196; as much
and as good, 82; empty original world, 21; end of commons, 84; labor theory of
value, 75, 83; productivity,
84; rejects waste, 80; resource depletion, 77;
spoilage limitation, 80; suffi-
ciency condition, 72

market failure, 10, 13, 99, 167,
168, 188, 196
Martin, Brian, 14
Mary, Queen of Scots, 49
Mazer v. Stein, 42
Microsoft, 59
Millar v. Taylor, 37
Milton, John, 183
Moglen, Eben, 130, 184
Mohamad, Mahatir, 161
monopolies, 47
*Moore v. Regents of the Universi-
ty of California*, 148
Moore, John, 148, 158, 160
More, Thomas, 115

National Institutes of Health, 99,
135
natural rights, 5, 10, 15, 17, 33,
38, 88, 109, 117, 154, 176,
177, 179
natural Rights,
as justification of IP, 10
NII White Paper, 6, 7, 62, 63, 106,
125, 133, 181
Nozick, Robert, 5, 19, 74, 75, 83,
166, 177-179

occupation, 73
Ockham, William of, 111, 112
Olivecrona, Karl, 83
Ovid, 110

Pardo, Arvid, 161, 162
patent trolls, 97, 165
patents, 28, 153, 154;
and developing countries,
102; as useful knowledge, 20;
basic vs. applied research,
99; computer programs, 58;
creation of, 47;
definition, 28; formalities,
29; inventive effort, 158;
medical procedures, 6, 148,
164; non-obviousness, 28;
on organisms, 155; origin of,
46; originality first required,
48; rarity of early patents, 39;
review by secretary of state,
39; strongest form of IP, 29;
term, 29; unpatentable subject matter, 28
Patterson, L. Ray, 2, 34, 35, 49,
125, 142, 147
Pettit, Phillip, 126, 128
Philistines, 45-47
Philistines:
monopoly on iron technology, 45
problem of consent , 81
problem of latecomers, 83
problem of scarcity, 82
property rights:
agglomerative tendencies,
194; not entailed by labor, 88
Prozac, 167

Public:
    as common world, 89
public goods, 18, 19
public goods problem, 9, 13, 130,
    167, 196
Pufendorf, Samuel, 3, 73, 75, 77,
    81, 82, 112; commons purely
    negative, 76; conquistador
    problem, 79; priority of pri-
    vate property, 77; Tully's cri-
    tique, 76

Rawls, John, 3, 18, 196
recordings:
    as performances, 56;
    as "texts", 26, 40, 51, 53;
    performances, 26
reverse engineering, 27
Rio Protocol On Biodiversity, 185
Rose, Carol, 159
Rose, Mark, 1, 2, 16, 26, 35, 142,
    153, 190
Rousseau, Jean-Jacques, 78, 80,
    93, 126, 181; conquistador
    problem, 79; constraints on
    appropriation, 79; not aware
    of tragedy of commons, 77

Samuels, Edward, 44, 134, 164,
    181
Sarony v. Burrow-Giles, 51, 54
Sartre, Jean-Paul, 89
Scotus, Duns, 111
Sen, Amartya, 128
Shiva, Vandana, 14
sleeping sickness, 167
Sreenivasn, Gopal, 70, 80, 84
Stallman, Richard, 119, 120, 122
Stationer's Company, 6, 33-38,
    49, 50, 125, 139, 202
Stationer's Copyright:

censorship, 49; critique by
    Commons (1694), 50; differ-
    ences from modern copy-
    right, 49
Statute of Anne, 26, 33-35, 38, 39,
    49, 125, 131, 181, 201
Stefik, Mark, 60
Stepansson, Kari, 100
Stevenson-Wydler Act, 168

Technology Transfer Act, 99
Tolkien, J. R. R., 92
trade secrets:
    definition, 29; differences
    from copyrights and patents,
    30
trademark:
    definition, 30; strong and
    weak marks, 30; term, 30
tragedy of the anticommons, 77
tragedy of the commons, 77
TRIPS agreement, 40
trusted systems, 7, 59, 60, 63, 125

*Universal Declaration on the Hu-
    man Genome and Human
    Rights*, 163
Utopia, 74, 115, 116

valuational solipsism, 90, 126
Ventner, Craig, 158

Waldron, Jeremy, 2, 127
Williams, Roger, 116
The Wind in the Willows, 190
Winstanley, Gerrard, 67, 110,
    117-120, 122
Winthrop, John, 92, 116, 117,
    119, 121, 122

Yen, Alfred, 92, 93, 106, 143-145

About The Author

Henry C. (Chett) Mitchell teaches philosophy and comparative religion at Indiana University South Bend. His research interests include early modern political theory, information ethics, theories of property, and philosophical aspects of cosmology. He received his doctorate in philosophy from the University of Notre Dame in 2002.